Studies in Computational Intelligence

Volume 743

Series editor

Janusz Kacprzyk, Polish Academy of Sciences, Warsaw, Poland
e-mail: kacprzyk@ibspan.waw.pl

The series "Studies in Computational Intelligence" (SCI) publishes new developments and advances in the various areas of computational intelligence—quickly and with a high quality. The intent is to cover the theory, applications, and design methods of computational intelligence, as embedded in the fields of engineering, computer science, physics and life sciences, as well as the methodologies behind them. The series contains monographs, lecture notes and edited volumes in computational intelligence spanning the areas of neural networks, connectionist systems, genetic algorithms, evolutionary computation, artificial intelligence, cellular automata, self-organizing systems, soft computing, fuzzy systems, and hybrid intelligent systems. Of particular value to both the contributors and the readership are the short publication timeframe and the world-wide distribution, which enable both wide and rapid dissemination of research output.

More information about this series at http://www.springer.com/series/7092

Tamiru Alemu Lemma

A Hybrid Approach for Power Plant Fault Diagnostics

 Springer

Tamiru Alemu Lemma
Department of Mechanical Engineering
Universiti Teknologi PETRONAS
Seri Iskandar, Perak
Malaysia

ISSN 1860-949X   ISSN 1860-9503   (electronic)
Studies in Computational Intelligence
ISBN 978-3-319-89112-5   ISBN 978-3-319-71871-2   (eBook)
https://doi.org/10.1007/978-3-319-71871-2

Printed on acid-free paper

This Springer imprint is published by Springer Nature
The registered company is Springer International Publishing AG
The registered company address is: Gewerbestrasse 11, 6330 Cham, Switzerland

To my family, siblings and parents

Preface

Gas power plants, which are common in distributed power generation, are complex machines, both in terms of design and method of operation. These have led to the need for specialized preventive maintenance (PM) procedures. One of the core elements of PM is intelligent fault detection and diagnostics. In the past, several fault detection and diagnostics algorithms have been developed using analytical, multivariate and soft computing methods. These algorithms—with different levels of complexity—have been applied to various designs of gas turbine power plants (e.g., one-shaft, two-shaft, three-shaft, or combined cycle arrangements). The challenge, however, is that even though several cases have been addressed using different metrics for performance evaluation, it is still short of meeting the expectations in the field. This book provides a hybrid approach to fault detection and diagnostics. Effectiveness of the methods is tested based on data acquired from actual cogeneration and cooling plant (CCP).

Models are developed for Normal Operating Conditions (NOC) by applying Neuro-Fuzzy (NF) method. For the dynamic case, the same approach is employed in the framework of Orthogonal Basis Functions (OBF). The fault detection is evaluated in the presence of adaptive model uncertainty calculations. For the bounded error case, Minimum Volume Outer Bounding Ellipsoid (MOBE) algorithm is introduced for the estimation of rule consequent parameters and model confidence intervals. Applications of the methods are demonstrated by considering a CCP having a 5.2 MW Gas Turbine, 12 Tons/hr Heat Recovery Steam Generator (HRSG) and 1250 RTH Steam Absorption Chiller (SAC).

Chapter 2 presents literature review focusing on general classification of fault detection and diagnosis techniques, model-based approaches, NF-based methods, and fault detection and diagnosis techniques as applied to a CCP. Chapter 3 concentrates on the theoretical background on nonlinear model identification. It also includes important aspects of nonlinear models and the effect of model order on model performance.

Chapter 4 focuses on the theory behind the calculation of model confidence intervals and the fundamental concepts in the construction of a model-based fault detection and diagnosis system. Furthermore, it elaborates on model uncertainty equations derived on the bases of two different assumptions: (i) identically and independently distributed error assumption; and (ii) bounded error assumption. Last section of the chapter explains the structure of a fault diagnosis system that relies on fuzzy method.

Chapter 5, in a step-by-step manner, shows how the theory elaborated in Chaps. 3 and 4 is applied to a CCP. Additionally, a new method is revealed that can be used to systematically account for the effect of change in operating modes and the effect of external factors to the accuracy of the models for normal operating conditions.

Application of the methods from Chaps. 3 to 5 is discussed in Chaps. 6 and 7. Rather than demonstrating implementation of the proposed method only, detail analysis is carried out for different magnitudes and types of faults, respectively. The analysis is performed by comparing performance of the NF approach against fault detection and diagnostics systems designed on the basis of principal component analysis and auto-associative neural network, respectively. Chapter 8 summarizes the findings and analysis and highlights on possible future research directions.

The aim of the book is to share the findings on power plant nonlinear model identification and fault diagnostics. Efforts have been made to provide detail analysis that can relate to practical application of the fault detection and diagnostics framework. The presented material is useful for researchers and practicing engineers.

Acknowledgements

I am thankful to Universiti Teknologi PETRONAS (UTP) for all the support provided for the research. Many people also positively impacted my work, either directly or indirectly. I would like to particularly recognize the technical support and mentorship given to me by Associate Professor Dr. Fakhruldin Mohd Hashim and Associate Professor Dr. Mohd Amin Majid. I always treasure the experience I gained by working with you. My last and highest acknowledgement goes to my lovely wife Tigabwa Yosef Ahmad and my son Mathias Tamiru, who have sacrificed a lot for me.

September 2017   Tamiru Alemu Lemma
                                                               Universiti Teknologi PETRONAS,
                                                                    Seri Iskandar, Malaysia

Contents

Chapter 1
Introduction

1.1 Overview

In general, fossil fuel based power producing plants may not stay longer at a performance level envisaged at the design stage due to aging, malfunction (e.g. guide vane drift, fouling, erosion, variable bleed valve failure), changing fuel composition, and changing operating conditions. Because of these, they often require preventive maintenance interventions. If the onset of abnormal conditions is not dealt with at early stage, it can lead to reduced performance, high green house gas emissions and system outage. With excessive NOx, CO, UHCs and CO2 emissions, it will be difficult to comply with the regulatory bodys requirements on emission control. The loss of productivity because of breakdown creates high financial penalties depending upon the duration of plant downtime. Failure to have a monitoring system also causes increased cognitive load on operators and inefficient use of maintenance staff time. Since the plant is featured by high energy throughput, failure to detect even small reduction in efficiency may lead to high financial loss.

It was reported that maintenance cost of a power producing plant may reach up to 30% of the total electricity generating cost [1]. According to the script by North East CHP Application Centre [2], for a typical gas turbine the cost may fall in the range of 0.008–0.012 $/kW-h-year. A similar range was documented by Onsite Sycom Energy Corporation (OSEC) [3]. For a combined cycle power plant, the maintenance cost is reported amounting to 17% of the plant life cycle cost [4], Fig. 1.1. For absorption chillers, this cost is reported as $18–$31 per ton of cooling depending upon the type of design, maintenance strategy and operating conditions. In light of the data, it can be said that any improvement on the performance of the existing maintenance practice leads to significant cost savings. In fact, a study by Rosen [5] has revealed that a saving of about 30% in maintenance cost can be achieved by simply shifting from preventive maintenance to Condition Based Maintenance (CBM) in which a Fault Detection and Diagnosis (FDD) system plays a main role. Improving the FDD system, therefore, would mean improving the CBM capacity.

© Springer International Publishing AG 2018
T.A. Lemma, *A Hybrid Approach for Power Plant Fault Diagnostics*,
Studies in Computational Intelligence 743,
https://doi.org/10.1007/978-3-319-71871-2_1

Fig. 1.1 Plant life cycle cost
for a combined cycle power
plant [5]

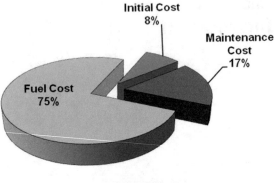

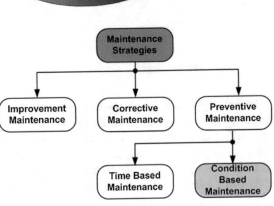

Fig. 1.2 Classification of
maintenance strategies

There are three types of maintenance strategies, Fig. 1.2. The first is Improvement
Maintenance (IM) that deals with maintenance considerations at the manufacturing
stage of the unit. The aim of this strategy is to do away with any maintenance requirement. But, because of the limitations on the material properties and manufacturing
method, parts are always designed for finite life. Hence, it is difficult to realize the
strategy practically. The second type is called Corrective or Reactive Maintenance
(CM). In this method, a part is replaced when it fails i.e. the approach is reactive
rather than proactive. It is appropriate when frequency of failure of a part is high. But,
it can cause unnecessary down time and high maintenance cost if the failed part is a
major part. The third type is known as Preventive Maintenance (PM). PM is divided
in to Time-Based Preventive Maintenance (TBM) and Condition Based Maintenance
(CBM). TBM is schedule or periodical attempt to prevent failure before it occurs.
Being a traditional approach it requires prediction of optimum time for maintenance.
Determination of the optimum time is always difficult. Unlike TBM, CBM also
called predictive maintenance is a proactive strategy in the sense that it recommends
action only based on existing conditions of the plant. It is advantageous as it reduces
unnecessary down time and maintenance cost.

The selection between maintenance strategies depends on the criticality of the unit.
Often, combinations of methods are used in one plant. But for plants as complex as
a PCP, CBM is a better choice for many reasons. Properly employed, CBM offers

the benefits of convenient maintenance scheduling. CBM is being used as a base for developing advanced maintenance approaches like Reliability Centred Maintenance (RCM), Intelligent Maintenance system and e-maintenance.

CBM involves three steps: data acquisition, data processing and decision making [6]. Data acquisition is the collection of relevant data required for maintenance. Data processing is the cleaning and analysis of different type of information collected in the first stage. Several methods are available for data analysis. But the main difficulty is the size of the information to be processed. The third stage is decision making. It entails fault detection, fault diagnostics and prognostics. The prime objective of this book is fault detection and diagnostics.

A fault in a power plant refers to any deviation in the current operation of the component, unit or system from expected or standard operation. It could occur for reasons other than regular changes caused by changing external factors e.g. temperature, pressure, humidity or change in fuel composition. The focus of fault diagnosis is to determine the kind, size, location and time of detection of faults. Fault prognosis, on the other hand, is targeted at the forecast of faults before they occur. For an effective CBM system both are equally important with the condition that the latter two cannot be done with out having the fault detection tool.

It is worth highlighting that, in CCPs, a combination of TBM and CBM are the current practice. In TBM, servicing of a system is performed according to the time table suggested by the manufacturers while CBM dictates things to be corrected right after they are detected and diagnosed. In contrast, with CBM early planning is possible, fast actions can be made, and resources can be used efficiently. Abnormal condition detection is the key step for the remaining actions. If the condition is a reduced performance, knowing it at the developing stage and the magnitude of the derangement relative to a normal operating state helps to estimate the time to reach to an unacceptable level [7]. If things like that are identified, the planning can be performed at early stage.

This book elaborates on the design of a fault detection and diagnostics system using hybrid approaches. A typical power plant consists of a gas turbine, boiler and steam turbines. In some cases there will also be a cooling system to enhance overall performance. The case considered throught the book is a cogeneration and cooling plant, which includes the three subsystems. For the actual system considered for case stsudy, the Distributed Control System (DCS) allows online monitoring of several parameters. As such, more than hundred parameters including pressures, temperatures and flow rates are available for trending. In cases where minor faults occur, the operators apply visual inspection to identify the causes. If things are not identified and the fault is a propagated fault or if it is safety related, they call for an expert from the vendor. This practice, however, creates delays in handling the problem and hence reduced availability of the system. In most cases, the practice included keeping solved cases in a way difficult to retrieve in full form. All these combined together calls for an enhanced fault detection and diagnosis system. It is important that the newly designed FDD system should be capable of using the available information and store analysis result.

1.1.1 Cogeneration and Cooling Plant in Brief

Cogeneration and Cooling Plants (CCPs) deliver electricity and chilled water utilizing a single source of energy that could be either from natural gas or liquid fuel. Typical design of a CCP is the one having a gas turbine as a prime mover, Fig. 1.3. The gas turbine could be a one shaft, two shaft or three shaft design. In the cogeneration section that is comprised of a Gas Turbine Generator (GTG) and a Heat Recovery Steam Generator (HRSG) the primary energy is used to generate electricity and convert boiler feed water to steam. In a tri-generation mode, the chilled water required for space cooling is thermally processed utilizing the steam from the HRSG. The connection between the HRSG and the Steam Absorption Chiller (SAC) is made through the Steam Header (SH). To cope with instant power and steam demand, the gas turbine is controlled by modulating the air and fuel flow rates. Variable inlet guide vanes (VIGVs) are used to control the air flow rates.

The CCP shown in Fig. 1.3 is a typical design. In the more advanced systems, units like inter cooler, regenerative heat exchanger, pre-heater, a system to capture CO_2, gasifier etc. are added to increase the overall efficiency and reduce green house gas emissions. In a combined cycle design, the steam from the HRSG is used to drive a steam turbine. There are also designs having the steam from the HRSG used for cooling the turbine nozzles and blades. In recent applications, the exhaust gas from the gas turbine is also seen directly used to drive an absorption chiller. All in all, Gas Turbine (GT) driven power plants are preferred to their counter part as a result

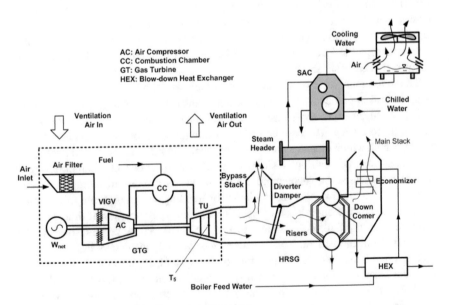

Fig. 1.3 Example of a cogeneration and cooling plant

of the many attractive features one of which are fast starting characteristics and the flexibility to use fuels other than natural gas.

1.1.2 Common Faults in a Cogeneration and Cooling Plant

A fault detection and diagnosis system must be designed to differentiate between actual faults caused by internal factors and changes caused by external factors. An example on the external factor causing performance derangement is change in air temperature and pressure at the inlet to the compressor. Change in the air temperature results an increase in compressor work and consequently reduces efficiency of the plant. This kind of change, however, is taken care by the control system itself. Internal factors comprise those intrinsic to the system and induced in the system. They are caused by component malfunctions, inadequate maintenance, aging of component parts, etc. Usually, these kinds of faults are covered by the control system. Fouling on the rotating and stationary blades of an axial compressor is a good example of an intrinsic type. Compressor wash is the common method to recover the performance from this kind of problem. Manufacturers suggest compressor wash to be carried out periodically- like in one to two month intervals. On site conditions, however, are the deciding factors. At times, compressor fouling and air filter clogging occur in a period less than what is mentioned by the vendor. Hence, continuous monitoring of the pressure on the suction side and compressor discharge pressure is very important and needs to be performed at all times.

Frozen measurement sensor is a major anomaly intrinsic type - in power plants. Failing to early notice faulty sensors creates repeated trips that may lead to reduced service life or fatigue failures of component parts. Performance deterioration of controllers is another fault type in gas turbines. Measurement sensors and controllers are very much sensitive to the operating environment. They deteriorate fast and become unstable in humid and high temperature environments unless special provisions are there to safeguard them from adverse conditions. The other fault in relation to sensors is wrong calibration. If the transmitter is in the feedback loop, the consequence of this is unexpected behaviour and it can lead to total outage. But, for transmitters outside the control loop the effect is minimal and is just on the performance evaluation of the system. For instance, wrong fuel flow reading leads to wrong efficiency calculation.

In the Gas Turbine Generator (GTG), fouling, erosion, corrosion and mechanical anomalies (e.g. VIGV drift) are the common performance related faults [5]. While erosion causes reduction in isentropic efficiency of the components, fouling and corrosion results in a reduced flow capacity at the inlet to the compressor and increased mass flow rate at the inlet to the turbine, respectively [8–11]. In fault detection and diagnosis algorithm testing, these faults are simulated by changing the mass flow rate, the pressure drop and efficiencies in a controlled range, [12, 13]. In terms of percentage of failures, the highest is linked to the hot sections of the gas turbine, Figs. 1.4 and 1.5.

Fig. 1.4 Major failures in
gas turbines [5]: capacity
less than 220 MW

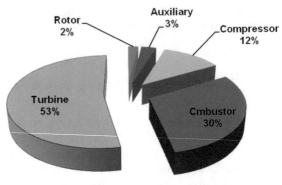

Fig. 1.5 Major failures in
gas turbines [5]: capacity
larger than 220 MW

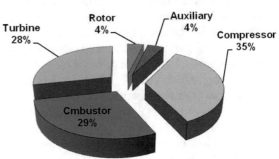

The HRSG has steam drum, water drum, water-walls, economizer, pre-heater and feed-water pumps as sub units. The prominent performance related faults common in the units include malfunction in the feed-water pump (damaged seals and erosion of impellers), tube leaks, and fouling in the remaining critical components. Fouling in the HRSG causes the exhaust gas exit temperature to increase, and exhaust gas pressure and steam production to decrease [14]. Regarding the flow rate transmitters and water level transmitter, correct calibration need to be achieved at the time. Minor error in the control setting can result in frequent HRSG trip. The exhaust gas entering to the HRSG is controlled by a diverter damper the opening position of which is manipulated by a hydraulic actuator. Fast response of this unit during operation is very critical for safety related reasons. Performance deterioration of the controllers and sensors in the control loop highly affects the whole system for the turbine is synchronized with the HRSG.

The SAC is featured by an Evaporator (EV), Absorber (AB), Heat Exchanger (HX) (low pressure and high pressure), pump (solution, refrigerant, and spray), expansion valve, Generators (GNs) and Condenser(CN). Fouling, air leaks and pump malfunctions are common problems to the adequate operation of the unit. Air leaks often cause crystallization. If not corrected, it would result in a decrease in the performance [15, 16]. Because the system works under vacuum pressure, air leaks at the joints may occur.

Performance related anomalies are difficult to detect by the control system. By design the control system has zero intelligence against performance related faults. It happened that the controller and measurement sensors are also susceptible to performance deterioration with time creating complex fault symptoms. The latter can lead to catastrophes and human fatalities if they are not detected at early stage.

CCP often operates under the surveillance of Supervisory Control and Data Acquisition (SCADA) system, Distributed Control System (DCS) or Power Management System (PMS). These systems provide plenty of signals for the plant operating personnel. They use the data to detect unusual operating behaviours. For a system driven by a single-shaft gas turbine, more than seventy signals are available for the gas turbine alone. Handling this data reach and information poor signals to make sense of it is expensive and beyond the capacity of any single person.

1.2 Research Gaps

Varieties of faults occur in a CCP. While timely maintenance response to reduce downtime creates cost savings, relying on humans to deal with any abnormal condition in an environment having large volume of data and complex operating process is often susceptible to wrong conclusions. It is, therefore, desirable to *develop a fault detection and diagnostics system that accommodates nonlinearity of the system and qualitative nature of diagnostic information naturally understandable by the operators.* Because of the high energy throughputs in CCPs, any attempt to enhance the existing maintenance strategy could lead to significant economic benefits.

CCPs are nonlinear systems attributed to several reasons (e.g. non-isentropic processes, operating point dependent input modulation, complex combustion sequence, auxiliary systems being connected to the turbine shaft, etc.). Hence, using linear models to characterise CCPs is far too short of representing the real behaviour. Conventional fault detection and diagnosis techniques Observers, parity relations and parameter estimation techniques — rely on detailed first principle nonlinear models (in fact, many of them use linearization over a certain region). However, it is often difficult to develop a first principle model. In some cases, a high-fidelity model may be readily available. But then, due to the repeated iterations involved in the property estimation and component matching, the computational time is far too high to be applicable for online use. Hence, one approach that could be used to remedy this problem is to run the model in an offline mode and save it as data driven model. The requirement, however, is that this approach also needs a suitable model. Soft computing or computational intelligence techniques are believed to be effective in this regard. This book will demonstrate application of Neuro-Fuzzy (NF) methods combined with data processing and other technologies available for model identification.

NF equires prior decisions on the number of inputs and the way the training and testing data are considered. Therefore, there is a need to partly on the combined use of thermodynamic, heat transfer and NF methods. To judge on the effectiveness of this proposal, it is necessary to assess the performance by making comparison against the likes of Principal Component Analysis (PCA) and Auto-Associative Neural Network (AANN).

Model confidence intervals are critical to the fault detection process. Having selected either Output Error (OE) model or Equation Error (EE) model as the preferred method, choices need to be made between identically and independently distributed (iid) error assumption, and bounded-error assumption in estimating the model parameters. Both are successfully applied in linear regression models. The extension of these to NF approach is to be elaborated further.

Past studies demonstrated that crisp residuals are directly considered as an input to a neural network model or a fuzzy relational model to create symptom-to-cause relationships. In another approach, first the crisp residuals are fuzzified based on heuristically defined membership functions and then models are developed in a similar way. The case of automatic generation of fuzzy membership functions from the fault detection procedure itself, especially for adaptive threshold based deigns, needs to be addressed.

1.3   Objectives to be Addressed

A reliable monitoring and fault diagnostics system is crucial for safe, efficient and uninterrupted operation of a power plant. Many approaches have been suggested as a means for efficient fault detection and diagnostics. Nevertheless, as would be revealed in the literature review, there is no unique method or approach that efficiently answers to real requirements — nonlinearity, uncertainty, incomplete experts knowledge etc. Driven by these facts, the book aims to develop and assess effectiveness of:

a. *Neuro-Fuzzy based Models for Fault Detection and Diagnostics:*
   A reliable model from first principles is required for a model based fault detection and isolation procedure. Once the model is there, it will be used to artificially generate output signal trends by inducing faulty input signals either in an online mode or off-line mode. For the latter case, things will be repeated and trends will be generated for as many abnormal cases as possible. Latter, NF training will be performed to save the cases for further call up while there are fault detection and isolation requirements. Because the model will be a complex one, using real time simulation of the model for the purpose of fault detection is hardly a first choice and will not be main goal of the book.
b. *Neuro-Fuzzy based models for the HRSG and SAC:*
   In the CCP, the heat recovery steam generator is one of the critical unit. During operation, a turbine may trip caused by excessive back pressure from the HRSG side. Diverter dampers part of the HRSG — are modulated to control the flow of exhaust gas to the HRSG. While the same element is manipulated to control the steam pressure, three element controls are used to adjust the feed water flow to the steam drum so that the water level is kept at the set point. The monitoring of this

unit also requires accurate models. In this case, however, controller settings and overall heat transfer coefficients are not available. The steam line includes also the steam absorption chiller. Steam flow rate is controlled to meet the chilled water outlet temperature setting. The overall heat transfer coefficients and the control setting are not known. The steps assumed for the gas turbine can be adopted to get an approximate HRSG and SAC models. For these systems, such a model will not be developed. Rather, NF identifications are applied based on actual data.

c. *Adaptive Threshold Calculation Algorithms, under Two Different Assumptions: (i) Normally and iid Error, and (ii) Bounded Error:*
   In fault detection and diagnosis, the calculation of threshold values is the difficult and critical step. If the threshold is greater than optimum value small faults will be missed and if the threshold is less than optimum value false alarms will be triggered. For acceptable performance, an adaptive threshold calculation technique must be applied — the main assumption here is analytical redundancy shall be used as a base approach.

d. *Algorithms to Link the Adaptive Threshold based Fault Detection Method with Fuzzy based Fault Diagnosis Algorithm:*
   In the adaptive threshold approach, the transition from fault detection to fault isolation is not an easy task for the threshold evolves with the dynamics of the system.

e. *Overall Plant Level Fault Detection and Diagnosis Strategy:*
   The CCP is three main units working in sequence. Each of them need an independent treatment but when it comes to overall monitoring an integrated scheme is what seems attractive. The health or unaccepted operation of one unit will directly or indirectly affect the other. It is then logical that, apart from developing a unit-by-unit fault diagnosis system, there must also be a plant-wide fault detection and diagnosis system.

In a nutshell, the books ultimate goal is to demonstrate the design of a fault diagnostics system using hybrid systems. The steps from literature review all the way until validation are covered.

References

1. Grber, U. (2004). Advanced maintenance strategies for power plant operatorsintroducing inter-plant life cycle management. *International Journal of Pressure Vessels and Piping, 81*, 861–865.
2. North East CHP Application Centre (2010) CHP Publications and Resources. http://www.northeastchp.org/nac/businesses/pubs.html
3. ONSITE SYCOM Energy Corporation (OSEC) (2000) The market and technical potential for combined heat and power in the commercial/Institutional sector. http://www.eere.energy.gov/de/pdfs/chp_comm_market_potential.pdf
4. Boyce, M. P. (2006). *Gas turbine engineering handbook.* UK: Gulf Professional Publishing.
5. Rosen, J. (1989). Power plant diagnostics go online. *Mechanical Engineering*

6. Jardine, A. K. S., Lin, D., & Banjevic, D. (2006). A review on machinery diagnostics and prognostics implementing condition-based maintenance. *Mechanical Systems and Signal Processing, 20*, 1483–1510.
7. Li, Y. G., & Nilkitsaranont, P. (2009). Gas turbine performance prognostic for condition-based maintenance. *Applied Energy, 86*, 2152–2161.
8. Kurz, R., & Brun, K. (2001). Degradation in gas turbine systems. *ASME Journal of Engineering for Gas Turbines and Power, 123*, 70–77.
9. Lakshminarasimha, A. N., Boyce, M. P., & Meher-Homji, C. B. (1994). Modeling and analysis of gas turbine performance deterioration. *Journal of Engineering for Gas Turbines and Power, 116*, 46–52.
10. Aker, G. F., & Saravanamuttoo, H. I. H. (1989). Predicting gas turbine performance degradation due to compressor fouling using computer simulation techniques. *Journal of Engineering for Gas Turbines and Power, 111*, 343–350.
11. Tabakoff, W., Lakshminarasimha, A. N., & Pasin, M. (1990). Simulation of compressor performance deterioration due to erosion. *Journal of Turbomachinery, 112*, 78–83.
12. Zwebek, A., & Pilidis, P. (2003). Degradation effects on combined cycle power plant performance-part I: Gas turbine cycle component degradation effects. *Journal of Engineering for Gas Turbines and Power, 125*, 651–657.
13. Toffolo, A. (2009). Fuzzy expert systems for the diagnosis of component and sensor faults in complex energy systems. *Journal of Energy Resources Technology, 131*, 1–10.
14. Port, R. D., & Herro, H. M. (1991). *The Nalco guide to boiler failure analysis*. New York: McGraw-Hill Professional.
15. Wang, S. K. (2000). *Handbook of air conditioning and refrigeration*. New York: McGraw-Hill, Inc.
16. Herold, K. E., Radermacher, R., & Klein, S. A. (1996). *Absorption chillers and heat pumps*. New York: CRC Press.

Chapter 2
Literature Review

2.1 Introduction

This chapter is intended to discuss the FDD methods in the area of power plants. However, apart from covering the methods applied to CCPs, it zooms into thermal systems in a wider perspective. The main benefits from this chapter is that it reveales possible research directions that reaffirm our choice of a specific method in Chap. 1. Further more, candidate areas for possible further research are also enumerated.

A power plant may experience abnormal or faulty state. This state could be a manifestation of a freezed sensor outside the control loop or a result of unrecoverable problem such as erosion of the turbine casing or the rotating blades. In some cases it could be a combination of two or more problems that happen to demonstrate similar symptoms, e.g. compressor fouling and VIGV drift. Regardless of the condition of sensors, however, we rely on sensor information to conduct fault detection and diagnostics. The presence of a sensor fault invalidates the calculation intended for optimization, supervisory control, and FDD. Hence, the ideal version of FDD system needs to take into account the expected conflicting situations, both in terms of fault combinations and availability of sensors for measurement and diagnostics.

The design of a fault detection and diagnostics system is affected by many factors: (i) detectability and isolability of the fault, (ii) the type of signals available for measurement, (iii) the state of the sensing element, (iv) the threshold assigned to the detection step, (v) richness of the diagnostic knowledge base, and (vi) adequacy of the inference system. Several approaches have been suggested to design a fault detection and diagnosis system.

© Springer International Publishing AG 2018
T.A. Lemma, *A Hybrid Approach for Power Plant Fault Diagnostics*,
Studies in Computational Intelligence 743,
https://doi.org/10.1007/978-3-319-71871-2_2

2.1.1 Basic Definitions

In this book, the meaning for key technical words and phrases related to FDD follow the definitions as provided by the SAFEPROCESS Technical Committee (International Federation for Automatic Control) [1].

Fault: *An un-permitted deviation of at least one characteristic property or parameter of the system from the acceptable/usual/standard condition.*
Failure: *A permanent interruption of a systems ability to perform a required function under specified operating conditions.*
Malfunction: *An intermittent irregularity in the fulfilment of a systems desired function.*
Fault detection: *Determination of the faults present in a system and the time of detection.*
Fault isolation: *Determination of the kind, location and time of detection of a fault. Follows fault detection.*
Fault diagnosis: *Determination of the kind, size, location and time of detection of a fault. Follows fault detection. Includes fault isolation and identification.*

2.1.2 Available Signals Versus Faults

Addressing the issue of fault categories in relation to the process model, signals accessible for system monitoring and time dependence of the faults themselves is worthwhile to the subsequent detail literature review. In connection to this, consider a controlled plant described by a block diagram given in Fig. 2.1. In general, a plant is characterised by input sensors, output sensors, controller, and actuators. The state of the plant is monitored by means of input signal $x(t)$, output signal $y(t)$, signals

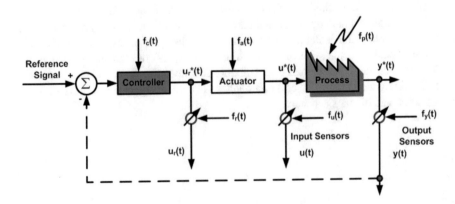

Fig. 2.1 A controlled system and fault topology

Table 2.1 Time dependence of faults

Types of fault	Relation	Examples
Abrupt	$f(t) = \hat{f}$, for $t \geq t_s$	Freeze sensor, FOD in compressors
Incipient	$f(t) = a_{11}(t - t_0) + a_{10}$	Fouling, Erosion and Corrosion
Intermittent	$f(t) = \hat{f}_i$ for $t = t_i$	Outliers

from the controller $u_r(t)$, and signals outside the control loop. The actual plant inputs $u^*(t)$ and outputs $y^*(t)$ are not directly available. The same is true for the controller output $u_r^*(t)$. Sensors are used to acquire these quantities. The feed back controller gets the output signal through an output sensor. In ideal case the output measured by these sensors are equal to the actual values. In reality, the measurements are affected by the characteristics of the sensing element and disturbances. Noise is the first thing that can not be avoided (true for output sensors). Besides that, for different reasons, e.g. stuck sensor, the output could be offset in a certain direction.

It can be inferred from Fig. 2.1 that a fault in the system can be of a controller fault $f_c(t)$, an actuator fault $f_a(t)$, a process fault $f_p(t)$ or a fault linked to the measurement sensors $f_u(t)$ or $f_y(t)$.

In relation to the process models, faults on inputs and outputs sensors, respectively, are often modelled either as additive type, $u(t) = u^*(t) + f_u(t)$, or multiplicative type, $f_u(t) = \delta u^*(t)$. Where the vector $f_u(t) = [f_{u1} \, f_{u2} \cdots f_{u.nu}]^T$ describes a specific fault signature and δ is the multiplier. Including the effect of measurement noise, $\tilde{u}(t)$ and $\tilde{y}(t)$, with the assumption that they are white, zero mean and uncorrelated Gaussian processes, faulty sensor signals can be modelled as

$$
\begin{cases}
u(t) = u^*(t) + \tilde{u}(t) + f_u(t) \\
y(t) = y^*(t) + \tilde{y}(t) + f_y(t)
\end{cases}
\tag{2.1}
$$

Additive faults manifest themselves as offsets of sensors where as multiplicative faults appear as parameter changes within the process – e.g. fouling in a heat exchanger or compressor. Faults are also distinguished based on their time dependence as abrupt fault (stepwise), incipient fault (drift like), and intermittent fault, Table 2.1.

So far we have focussed on those signals involved in the control loop only. In real application, there are also signals outside the control loop and used for the purpose of condition monitoring. A good example is to consider an industrial gas turbine, Fig. 2.2. The pressure drop on the suction side of the compressor is monitored to make sure that the air filter at the inlet of the duct is not dust clogged excessively. The other signals available for measurement range from lube system temperatures and pressures, turbine enclosure temperature, vibration signals at the shaft supporting bearings, generator winding temperatures, and acoustic signals to lube oil sampling.

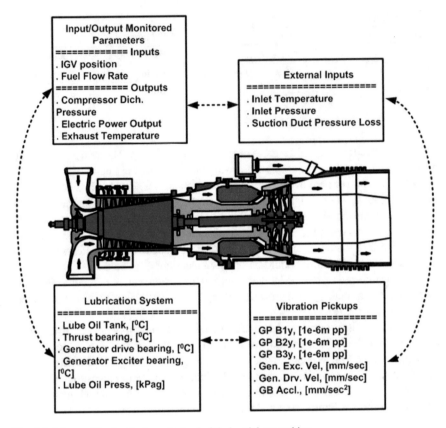

Fig. 2.2 Measurable signals for a single-shaft industrial gas turbine

Heuristic information is another source of monitoring signal. Human observations in the form of noises, colours and smells contribute to this group. Past maintenance history is also another source of heuristic information.

2.1.3 Classification of Fault Detection and Diagnosis (FDD) Methods

The classification of the fault detection and diagnosis techniques follows the signals available for measurement, the techniques used for signal processing and the type of inference system. Literature reviews on the different approaches are available in conference papers [2, 3], journal papers [1–12], and books [13, 14].

The first grouping for FDD is as fault detection and fault diagnosis, Table 2.2 and Fig. 2.3. Detailing the fault detection a little bit, the general classification emerges as model-free and model-based. While model-based approaches rely on a mathematical

Table 2.2 Time dependence of faults

Objective	Method	Examples
Fault diagnosis	Classification based	• Statistical based
		• Pattern recognition
		• Neural networks
		• Fuzzy logic
	Reasoning based	• Forward reasoning
		• Backward reasoning
Fault detection	Model free	• Hardware redundancy
		• Signal based
	Model based	• Quantitative model based
		• Qualitative model based

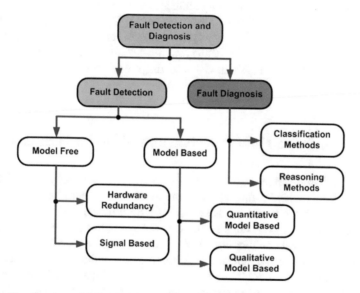

Fig. 2.3 Classifications of fault detection and diagnostics methods

model developed either from fundamental laws or based on measured input-output data, model-free approaches often make use of direct analysis of the collected signals.

There are several methods as model-free approach. These include physical redundancy, special sensors, limit checking, spectral analysis [15] and logical reasoning. In hardware redundancy design [16, 17], multiple sensors are used to measure a variable and voting techniques are used to decide on the faulty measurement. The drawback to this method is the extra cost to the system. Limit checking [17, 18] on the other hand works by comparing the real time data with the threshold that is preset by the operator. Though it is simple, it is not accurate enough for the thresholds are calculated not taking into account the effect of other variables in the system.

In fact, this makes it identical to a univariate statistical technique. Limit checking lacks sensitivity and robustness as it ignores spatial and serial correlations among the observations and variables. Some signals could have special frequency spectrum in normal and faulty conditions. In that case, spectral analysis can be used for fault detection and diagnosis. Logical reasoning takes the symptoms from the other methods and applies structured approaches in an expert style or without the involvement of the human operator.

A model based fault detection system doesnt need extra equipment or there is no voting involved. The method simply works making use of the discrepancy between actual output and model output of the system. The model can be developed employing either qualitative techniques or quantitative methods, Table 2.2.

Fault diagnosis is the determination of the time of detection, location, type and size of a fault. The available signals are translated into suitable symptom vectors and classification or diagnostic reasoning strategies are applied to reach to diagnostic decisions. Classification methods statistical based [19], pattern recognition methods [20], neural networks [21], and fuzzy systems [22–24] are used in cases where there are no available knowledge about fault-symptom causalities. Most of the time partial information is known for a system. If it is so, diagnostics reasoning strategies play a major role.

2.2 Model-Based Fault Detection and Diagnosis Techniques

The general classification on the model based approaches towards fault detection is given in Fig. 2.3. The main step in all the cases is the generation of fault symptoms by comparing a reference or normal operation model with actual data. The difference in the approaches lies on the way the symptoms are generated. But, all use models developed in either online or off-line mode. The model based approach is broadly classified as qualitative model based and quantitative model based. The general structure of a model based fault detection and diagnosis system is given in Fig. 2.4. To make diagnostic decisions possible, a residual $r(t) = y(t) - \hat{y}(t)$ featured by an offset from the threshold $th(t)$ is further processed by a classifier or a diagnostic reasoning strategy (Fig. 2.5).

2.2.1 Quantitative Model Based Methods

Analytical Methods (Observers, Parity Relations, Parameter Estimation Techniques, and Kalman Filters)

Models in this group use mathematical representations of the system derived from first principle concepts and are applicable to information rich systems, mostly linear. They are often referred to as analytical based approaches. Output observer

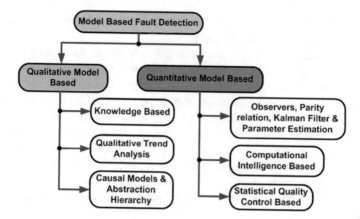

Fig. 2.4   Classifications of model based fault detection and diagnostics methods

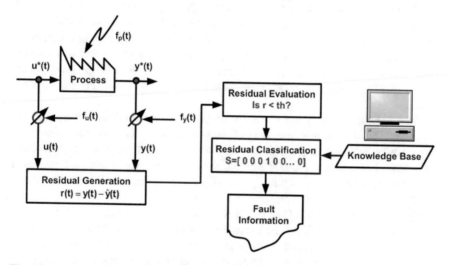

Fig. 2.5   Structure of a model based fault detection and diagnosis system

(Luenberger observers) [25, 26], and Unknown Input Observer (UIO) based fault detection methods make up the first group. The design structure corresponding to the two methods features each output modelled as a function of all the inputs and each input modelled as functions of all inputs and outputs except the input under consideration.

In the dedicated observer scheme, the output observers are designed for each output sensor assuming that the state of the system is observable from the measured quantities. Changes in the estimated states are used as fault indicators. The necessary condition for this approach is that the observers need to be designed in such a way that they converge to zero under normal operating conditions. The drawback in this approach is the requirement for a mathematical model and the need for a number of observers as there are number of outputs.

The first use of observers for fault detection and diagnosis is dated back to the early 1970s. Beard [27] at Massachusetts Institute of Technology (MIT) was the first to apply observer-based fault detection models. Since then different observers have been applied to sensor fault detection of steam generators [28], power plants [29], turbofan engines [30], heat exchangers [31, 32] and coal mills [33].

One difficult task prior to designing the decoupled observers is the calculation of system matrices. Simani, Fantuzzi et al. [34] investigated the used of model identification techniques (ARX, and EIV models) to obtain the system matrices in the design of input observers and output observers for an industrial gas turbine. The data used for the identification process is obtained from simulation of a MatLab/simulink model of the system.

Belonging to the same group of model based approaches is parity equations [35–38]. In this case, while the model is developed applying first principles, the model is simulated in parallel with the actual system and a residuals are calculated as difference between the model outputs and outputs from the plant. Through a comparison of the magnitude of the residuals to preset thresholds, the state of the system is identified. Good side of the parity approach is simplicity. However, for a Single Input Single Output (SISO) system, the residual will not be enough to isolate between different faults.

The other class of analytical model based approach for fault detection is parameter estimation [39, 40]. In this case, the technique is based on the hypothesis that faults may be reflected in the system parameters. For an actual system, physical parameters are seldom known. In parameter estimation method, the parameters are estimated from measured input and output signals, $u(t)$ and $y(t)$. In real time application, it mostly uses Recursive Least Squares (RLS) method.

De la Fuente and Vega [41] tested RLS and neural networks for fault detection in a water treatment plant. Belle and Isermann [42] applied RLS and local linear fuzzy models to detect and diagnose faults in an industrial scale thermal plant. The fuzzy models are used to capture the parameters during normal operation of the system. Abidin, Yusof et al. [43] used RLS with fuzzy inference system to diagnose faults in DC motors. Other related investigations using this approach include research works by Weyer, Szederknyi et al. [44], Kumar, Sinha et al. [45], and Yoshida, Kumar et al. [46].

The last approach in this category is Kalman Filter (KF) technique. In this case, structure of the detection device is similar to output observers. The major difference id, KF recursively estimates model matrices assuming that the measurement noise is white and Gaussian.

KF based observer design has been used by Tylee [47], Usoro [48], Li [49], Hines [50], Simani [51, 52], Zhang [53], and Ryerson [54]. KF is mostly applied for signals whose noise to measured signal ratio, $\|y^*(t)\|^2/\|\tilde{y}(t)\|^2$ and $\|u^*(t)\|^2/\|\tilde{u}(t)\|^2$, are relatively low.

In summary, all the analytical approaches are mostly applicable to systems with limited number of inputs and outputs, in fact to linear systems. For large scale systems, it is difficult to apply the methods given the cross-coupling between system variables. With analytical models it is also difficult to accurately consider the effect of model uncertainties caused by disturbances and noise. Robust fault detection and diagnosis strategies could improve the latter effect but even then it would be at the expense of delayed result in the detection of incipient faults.

Multivariate Statistical Methods

Multivariate statistical methods are qualitative model based techniques. All of them in this category are derived directly from measured process data. In this class, we find Principal Component Analysis (PCA), Partial Least Squares (PLS), Fisher Discriminate Analysis (FDA), and Canonical Variate Analysis (CVA). In the following, we only consider the PCA method.

In PCA approach, the data is mapped to a reduced dimension space making use of Singular Value Decomposition (SVD) on the covariance matrix defined by the input-output data. If the data matrix is $\mathbf{Z} \in R^{N_d \times (n_x + n_y)}$ then, when written in terms of parameters from the projected or lower dimensional space

$$\mathbf{Z} = \sum_{i=1}^{k} \mathbf{t}_i \emptyset_i^T + \mathbf{E} \tag{2.2}$$

where, $\mathbf{E}$ is the residual matrix, $\emptyset_i$ is the loading vector eigenvectors of the covariance matrix $\mathbf{Z}^T \mathbf{Z}/(N_d - 1)$ and $\mathbf{t}_i$ is the projection of $\mathbf{Z}$ in the direction of $\emptyset_i$. N_d, n_x and n_y are number of data, number of inputs and number of outputs, respectively. The number of loading vectors k considered for the projection is determined either by proportion of trace explained method or SCREE test [55]. For fault detection using PCA, Hetellings T^2 and Q-statistic are used as condition measuring criteria while contribution plots are used to isolate the variables responsible for the fault. Extensions of the classical PCA method are available in different forms: Moving PCA (MPCA), adaptive PCA, exponential weighted PCA [56]. The application of PCA for fault detection and diagnosis can be found in the area of air handling units [57, 58], centrifugal chillers [59–61], steam generator of a nuclear reactor [61, 62], gas turbines [64, 65], and Heating Ventilation and Air-Conditioning (HVAC) [19, 66, 67]. In Xu, Xiao et al. 2008 [61], the PCA based fault detection is enhanced by pre-processing the measured data by Discrete Wavelet Transform (DWT). PCA is also combined with other approaches like Hidden Markov Chains [68], and neural networks [69] to exploit dimension reduction capability of PCA. For a comparison between PCA and Adaptive Neuro-Fuzzy Inference System (ANFIS), refer to the paper by Maghsooloo, Khosravi et al. [70]. It demonstrated how PCA out performed ANFIS in terms of fault detection ability.

One advantage of this group of methods is that they dont need first principle model. Instead, they only rely on the input-output data. The other advantage is that they provide dimension reduction. The drawback of PCA, however, is that it assumes the projected data follows Gaussian distribution, which is rarely true. Besides, it doesnt take into account serial or self correlation. In fact, this property is shared by PSA as well. Some research works suggested Independent Component Analysis (ICA) to deal with the non-Gaussian condition, [71]. The next section will focus on computational intelligence based fault detection techniques.

Computational Intelligence Based Methods

CCPs reside to nonlinear systems. When the effect of measurement noise is significant and the system is featured by uncertain operating parameters, FDD users rely on computational intelligence techniques: neural networks, fuzzy logic and metaheuristics.

Neural networks are computing elements inspired by the behaviour of nerve cells in the brain. They are well known for their function approximation and classification abilities. A neural network having one hidden layer is capable of approximating any nonlinear function [72]. They are also chosen over the other approaches for their characteristics to generalize from a given training data set, tolerance to measurement noise and fast execution time once they are trained. The drawback of a neural network, however, is lack of transparency of the resulting model in human understandable terms.

Reviewing the application of CI methods to fault detection and diagnosis were the subject of many research [3]. Neural networks are used in two ways regarding process fault detection and diagnosis in power plants. The first use is as residual generator [21, 73–77] in the sense that neural networks are employed to capture Normal Operating Condition (NOC) of the system. The second application is as fault isolator or classification technique [17, 51, 78–82]. Some researchers also applied neural network as One Step Diagnostician (OSD) [83–87]. In this case, detection and isolation of faults are performed from available measurements without the need for prior generation of intermediate signals as residuals. In the report by Chun-ling [86], Radial Basis Function (RBF) networks trained by Genetic Algorithms (GAs) are used to formulate the OSD for such units as steam generator, steam turbine and condenser. Group Method of Data Handling (GMDH) neural network, [63, 88–91], is one design from the same class used for developing a fault detection and diagnosis system.

Neural networks are also used together with other methods. Sreedhar et al. [92] investigated the use of neural networks and sliding observers for fault detection in a thermal power plant. In some cases PCAs are also used for dimension reduction in the process of model identification through neural networks [69, 93]. A research work by [94] addressed the use of Kohonen Self-Organizing Maps (SOM) to reduce the training data size prior to using the data for formulating the neural network model. Simani, Fantuzzi et al. [95] and Simani [96] used Kalman filters and neural networks for fault detection and diagnosis, respectively, in an industrial gas turbine. Leger, Garland et al. [17] examined the feasibility of using neural networks combined with statistical control charts (CUSUM) for fault detection and diagnosis. Pattern recognition techniques are also combined with neural networks in developing dedicated models for different operating regions of a power plant [97]. Neural networks are also integrated with expert systems in the work of [98]. While the neural network is used for fault detection, the expert system is applied for fault diagnosis.

One special type of neural networks is the Auto-Associative Neural Network (AANN). First proposed by Kramer [99], the network has five layers with the middle layer called bottleneck layer. The network is suitable to formulate nonlinear principal component analysis on a given data set. Other than image processing, the network has been used by some researchers for sensor fault detection in gas turbines [100–102].

Fuzzy logic is another class of computational intelligence group. In fuzzy logic, the classical crisp description of concepts is generalized to fuzzy sets whose membership now assuming partial belongingness or full membership to a given set. Fuzzy logic was designed originally to describe vague linguistic concepts. A good example is temperature high, very high, low, very low, medium etc. As compared to neural networks, fuzzy approach is more appropriate for fault isolation. Fuzzy logic allows integration of human knowledge into the diagnosis process. One drawback of fuzzy modelling is its complexity and time consuming modelling procedure.

Similar to the neural networks, fuzzy systems can be used either as residual generator or pattern recognition technique. The latter use of fuzzy systems is common for cases where there is no prior knowledge about fault-symptom causal relationship in the design of a fault detection and diagnosis system. Some of the fault detection and diagnosis approaches exploiting fuzzy concepts include works by Diao [103] and Ogaji [104] both in the area of gas turbines.

Observers combined with fuzzy logic, in a serial way, are used for fault detection and diagnosis in high pressure pre-heaters [105]. The observers are applied to generate the residuals while the fuzzy logic is used to process the residual for fault isolation reasons. Upadhyaya, Zhao et al. [63], and Zhao and Upadhyaya [106] examined the use of Adaptive Fuzzy Inference System (ANFIS) for fault detection in a nuclear power plant.

Evolutionary computing techniques genetic algorithm, genetic programming and evolutionary strategies are derivative free global optimization tools. In fault detection and diagnosis design, they are used to estimate neural network weights [86] and fuzzy model parameters. Guo and Uhrig [94] applied genetic algorithm searching strategy to determine optimum number of inputs to a neural network model. In their work, SOM was adopted to reduce the data size.

Computational intelligence techniques, when they are implemented as stand alone design, cannot efficiently model complex systems even if they provide better performance as compared to first principle modelling approaches. Distributed fault diagnosis system design is the suggested approach to alleviate the drawback. One such design was suggested in Koppen-Seliger, Kiupel et al. [105]. They reported residuals considered all together during fault isolation only. Multi-agent based fault diagnosis design that uses neural networks for modelling sub-systems of a complex plant is discussed by Heo and Lee [107]. A similar approach is also proposed by Arranz, Cruz et al. [108]. Hierarchical and decentralised neural network based approach is investigated by Ogaji and Singh [109]. A two step decentralized approach using Petri nets for the first steps and neural networks for the second is also suggested by Power and Bahri [110]. The neural networks are used to indicate the exact location of a fault in a certain section of the power plant.

Neural networks have learning abilities while fuzzy systems are capable of representing knowledge as sets of if-then rules. The combined use of neural networks and fuzzy systems possesses the advantages of both. There are several ways to combine neural networks and fuzzy systems [111–117]. One way is to create a fuzzy neural model using fuzzification block to map the measured data to a different scale and feed the result to a neural network system [118]. In the other design, Self-organising Maps (SOM) can be used to pre-process data, e.g. clustering or noise removal, for fuzzy systems. In the other design the fuzzy model can be adopted in the neural network allowing the parameters to be estimated using neural network learning algorithms.

Palade, Patton et al. [111] used NF networks for fault diagnosis in an industrial gas turbine. They used simulated data generated with the inlet guide vane position and fuel flow rate considered as inputs. Actuator faults and compressor contamination faults are among the cases considered in their study. NF networks representing Mamdani models were used for fault isolation. Zio and Gola [119] investigated the use of NF approaches to fault classification on the gland seals of pumps of the primary heat transport system in a nuclear reactor.

Uppal, Patton et al. [120] applied NF approach to develop a fault diagnosis system for a process featured by multiple operating points. For assumed number of faults and operating points, the method suggests number of decoupled observers extracted from the NF models. The method was tested using the data from DAMADICS valve benchmark problem [121]. Robust fault detection scheme based on NF networks was developed by Korbicz and Kowal [122]. They used bounded-error assumption to define model confidence intervals. The approach was tested by experimental data for a flow control valve. Tan, Rao et al. [123] presented the combined use of fuzzy systems and Adaptive Resonance Theory (ART) networks to the detection and diagnosis of faults in the cooling system of a power generation plant. The same system was previously studied by Chen, Lim et al. [124] applying fuzzy neural networks integrated with rule extraction scheme.

Razavi-Far et al. [125] investigated the use of NF networks for fault diagnosis in a U-Tube Steam Generator (UTSG) that is part of a nuclear power plant. They trained the models based on a data from UTSG simulator. The NF for residual generation was trained by a Locally Linear Model Tree (LOLIMOT) algorithm while the NF model Mamdani type for fault isolation was trained by Genetic Algorithm (GA). Applications of ANFIS to model identification and fault isolation can be found in Zhao and Upadhyaya [106]. NF approaches as applied to pattern classification for the purpose of fault isolation is presented by [119].

In the other use of combined approaches, Evsukoff and Gentil [126] have tested neural network models feed forward and recurrent designs whose inputs are fuzzified simulated data. The resulting models have been used for fault detection and isolation in nuclear reactors.

2.2.2 *Qualitative Model Based Methods*

The group in this section entails knowledge based systems (cause effect graphs and expert systems), Qualitative Trend Analysis (QTA), and causal models and abstraction hierarchy.

The expert system can be of a shallow-knowledge expert system, deep-knowledge expert system, or a combination. In the shallow-knowledge approach, the algorithms are developed based on knowledge from the experience of humans. The knowledge in the expert system is represented using one of such techniques as predicate calculus, production- rules, frames, scripts or semantic networks. As a searching strategy, all Artificial Intelligence (AI) techniques use forward chaining or backward chaining scheme. For a large searching space, the depth-first and breadth-first searches are applied.

Expert systems have been applied to boiler feed water systems [127], acetic acid extraction process [128], turbo-generators [129], and refrigeration process of a hydraulic power plant [130]. The drawback in these methods is the difficulty to apply the methods to large scale systems for they require large amount of effort. Often, the design is customized to specific application.

Causal analysis is based on fault-symptom relationships. Signed Directed Graph (SDG) is one approach that uses causal relationships. Symptom-Tree Model (STM) also resides to the same group. SDG has been applied to air-conditioning [131], and de-aerator system of a power plant [132, 133].

Model Based Diagnosis (MBD) that is based on logical statements and relying on detail modelling of the system also belongs to this group.

2.3 Fault Detection and Diagnosis Techniques as Applied to CCP

Conroy et al. [134] used symptom tree approach for fault diagnosis. Shallow knowledge systems and deep level approach was also tested. The case considered was a combined heat and power unit. A monitoring and diagnostic approach based on multiple aspect modelling and model interpretation is described in [135] by considering a cogeneration plant.

In [136], they discussed a real-time fault diagnosis system for a cogeneration plant. They relied on hierarchical fault propagation models to deal with non-interacting multiple faults. The nodes at each level are attached to causal relations describing failure modes between subsystems. The approach includes fault propagation probabilities and fault propagation time-intervals. Their method has resemblance with the works of Convroy [134] and Sztipanovits [135] for they used multiple-aspect modelling approach Hierarchical Process Model (HPM), Hierarchical Component Model (HCM) and Hierarchical Fault Model (HFM) that takes into account the component structure and functional structure of the plant.

Kumamaru, Utsunomiya et al. [137] has carried out fault diagnosis in district heating and cooling plant. Used were cause-effect tree diagrams to find the causes of faults e.g. reduced capability, pressure malfunction etc. and an analytical method for judging the malfunction rate. It handled a system comprised of vapour compression chiller, thermal storage tank and the associated pump network. Interesting part of their work is that dialogue style expert system is employed for fault diagnosis and the technique of adaptive adjustment of the threshold level is applied to realize sensitivity in early detection of operation anomalies. However, with the method being ad-hoc, limited numbers of faults are considered and the case of multiple faults is not considered at all.

Perryman and Perrott [138] and Perryman [139] reported on an intelligent control and monitoring system developed for a stand-alone combined heat and power plant. Artificial neural network is used to describe Normal Operating Conditions (NOCs). A residual calculated between the actual operation data and the NOC model output is used to detect if the performance is going to the unwanted region. In their extended work, they have also used Multilayer Perceptron (MLP) to normalize the operation data before it is feed to the residual calculation block.

Renders et al. [140] investigated the application of Radial Basis Function (RBF) networks to a nuclear power plant. Data from simulation model of the plant is used to train the neural networks. Input variables to the neural network are decided based on first principle concepts and applying such criteria as availability, completeness, minimality and sensitivity of the state variable to faults. Guglielmi et al. [79] tested the effectiveness of using multilayer feed forward and radial basis function networks in the heater section of a power plant. Data generated by a simulated model of the plant was used to train and test the approaches. Leaks in the feed water pipes, malfunction of a draining valve and malfunction of a sensor are the cases considered in their study. Batanov and Cheng [141] developed fault diagnosis expert system for ethylene distillation plant; they used shallow knowledge and deep knowledge of the process. Their work and Convroy et al. [134] work resemble each other as they use qualitative model based approaches (see Sect. 2.2.2).

Bonarini and Sossaroli [142] studied fault diagnosis in a steam-turbine driven plant based on opportunistic models models with different levels of abstraction. Munoz and Sanz-Bobi [143] developed incipient fault detection system based on probabilistic radial basis function approach. They considered condenser of a power plant as a case study. In their work degree of significance of the residual signal is estimated based on the input data adaptive residual calculation. As an optimization tool, they used low-memory quasi-Newton method.

Abu-el-zeet and V.C. Patel [144] developed a data based power plant condition monitoring system using K-means clustering, Euclidian distance calculation and Vector Quantization.

Biagetti and Sciubba [145] depicts a description of an expert system capable of monitoring the performance of a cogeneration plant. They developed performance indicators based on the real-time data. A hybrid semi-quantitative monitoring and diagnostics system for the same system is given in [146]. Though the approaches are simple both of them consider limited number of faults and the cases of multiple faults are not included.

In a different model based design, Valero, Correas et al. [147] and Lazzaretto and Toffolo [148] have used thermo-economic and exergetic approaches to detect and diagnose faults in a Combined Heat and Power (CHP) plant. The techniques describe operation anomalies in terms of additional irreversibility. Akin to the other methods, though, their approach requires a reference model and actual operation data.

Thomson et al. [149] showed the use of Statistical Quality Control (SQC) to fouling detection in a CHP plant. Their method is designed to work with flow rate measurements as an input. Prediction of heat transfer coefficients by Exponentially Weighted Moving Average (EWMA) method with suitable control charts allowed them to successfully monitor the plant.

Flynn et al. [150] focused on a Combined Cycle Gas Turbine (CCGT) plant and investigated the use of PCA and PLS for fault detection and diagnosis. The two tools are applied for monitoring the system with the PLS extended by incorporating neural networks. The neural network is incorporated to cover nonlinear operating conditions. Structure wise, they applied multi-block approach to effectively handle critical components of the plant. Niu et al. [151] developed a reformative PCA-based fault detection method for a thermal power plant. K-means clustering (for classifying the data under different operating regions) and fuzzy partitions are also used to supplement the method. Odgaard et al. [33] considered power plant coal mills fault blocked inlet pipe. They compared the optimal input observer approach with a data-driven approach based on dynamic PCA and PLS. In the end, they proposed a hybrid model data driven methods used to drive the optimum unknown input observers.

Odgaard and Mataji [152] used optimal unknown input observer approach to detect faults in a coal mill. The linearized model is developed from a nonlinear model of the mill. Hines, Don et al. [153] performed a study on analytical redundancy and neural network based Fault Detection and Isolation (FDI) system for a nuclear power plant. Their work was targeted at solving the problem of formulating a single FDI system covering the whole system and avoiding the problems coming due to coupling problems. Multilayer feed forward neural network with three alternative training algorithms (back-propagation with momentum, conjugate gradient method and Levenburg Marquardt algorithm) were tested.

Korodi and Dragomir [154] developed a mobile fault detection and diagnosis module. The case considered was a Geothermal Power Plant (GPP). Correlations derived from simulation results of the mathematical model for GPP is the reference to formulate the strategies. Their study focused on only on the stationary operating regions.

Fast and Palme [155] recently applied ANN to condition monitoring of CHP plants. While they use the method to capture NOCs, they included the calculation of production cost. According to their work, they have integrated it in a Graphical User Interface (GUI) with the option of linking to the CHP computer system. In fact the addition of GUI put their work to the same level with the work of Abu-el-zeet, Z. H. and V. C. Patel [144], and Arranz et al. [108]. Except the addition of cost calculation and GUI, the method is similar to the work of Szczepaniak [74].

2.4 Summary

The purpose of this chapter has been to explore the FDD technologies as applied to thermal systems in general and to CCP in particular. While considering different techniques in terms of design structure, advantages and drawbacks it was our main objective to reveal clear research directions and possible areas where new contributions could be made. The outcomes are summaried as follows:

In general, a CCP involves integrated operation of auxiliary systems and varieties of signals available for measurement. While there is a need for high fidelity model to investigate how the system responds to different operating conditions, the development of such a model is complicated by availability of limited design point data. In most of state of the art FDD methods, quantitative model-based techniques designed relying on historical data are frequently used. Even then, rigorous treatment of gas-path, lube system, generator coils are hardly available. The same is true with the combined handling of GTG, HRSG and SAC.

Nonlinear relations govern the mechanism of energy exchange in the CCP. While the uses of analytical methods are often ruled out for such systems, we have witnessed few cases where linearization around a certain region is used to formulate representative linear models. As compared to analytical methods, multivariate statistical methods gained wide spread application. Nevertheless, the latter method overlooks serial correlation, which makes it not flexible enough to deal with variable operating conditions. In large scale and dynamic systems, the use of PCA is mostly seen as dimension reduction technique.

The review has also revealed that there is hardly a single method capable of demonstrating all the characteristics needed for designing a succesful fault detection and diagnosis system for a CCP. Many of the available approaches tend to apply hybrid techniques. In terms of accommodating nonlinearity of the processes, tolerance against uncertainty caused by disturbances and measurement noise, and ability to represent knowledge in human understandable forms, hybrids of neural networks and fuzzy systems seems to be gaining more attention. However, the use of the hybrid approach in context of multiple auxiliary systems functioning together is not investigated well yet.

Models based on the use of Computational Intelligence (CI) techniques have been found falling in one of the three basic structures, Fig. 2.6. The first method is using either the neural network or the fuzzy system used as a classification tool, Fig. 2.6a. In this case, all the input-output data are feed to the model and fault index delivered as an output. While this method works for simple systems with limited number of output features, it tends to be troublesome as the parameter size increases and if the process is featured by multiple operating conditions. In the second design, the intelligent model is still used as a classifier but the inputs are calculated by another algorithm, Fig. 2.6b. In this case, the numbers of inputs are limited to the number of states being monitored. A good example is the research work by Simani, Fantuzzi et al. [95]. The last approach is to use the intelligent system as modelling technique, Fig. 2.6c. In all the cases, curse of dimensionality seems an issue that is yet to be solved.

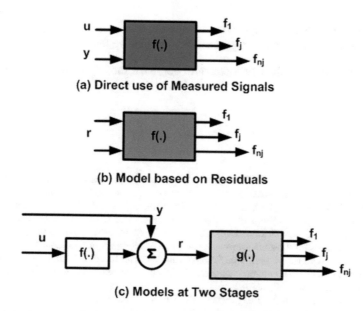

Fig. 2.6 Basic application oriented structures of CI based FDD

In dealing with a power plant featured by multiple operating regions, two approaches have been suggested. One is two develop dedicated neural network based models for each region [97] and the other is developing a neuro-fuzzy model covering the whole region but the result of NF model converted into a number of observers as there are operating regions [120]. Another approach that may go with the aforesaid points is reducing the size of a training data using clustering techniques. In this case cluster centres, rather than the real data, are used as an input to the neural network [94].

One important outcome of the literature review is that, in all the models featured by the prediction of current output based on past inputs and past outputs, the model order is decided by minimizing Euclidean norm of the modeling error using optimization algorithms. This, however, while resulting in fairly accurate prediction, may not guarantee model parsimony. Recently, the use of Orthonormal Basis Functions (OBFs) in model predictive control showed that parsimonious models could be developed if the models are constructed in the framework of OBFs. The use of OBF in the design of FDD systems is, therefore, one area that deserves detail study.

Finally, in light of the findings discussed so far, while designing the planned FDD system, we intend to concentrate on the following key areas:

- Dynamic models in the framework of GOBFs, CCP
- Multi-agent based design of the FDD system with a scope that covers main components of the,
- Semi-empirical models for the GTG and HRSG that could be used to generate data under implanted faults,

- Issue of multiple operating conditions, and
- Integrating fuzzy diagnostics procedure with adaptive fault detector.

In the next chapter, the development of nonlinear models using CI techniques is presented.

References

1. Isermann, R., & Ball, P. (1997). Trends in the application of model-based fault detection and diagnosis of technical processes. *Control Engineering Practice, 5*, 709–719.
2. Aishe, S., Weimin, C., Peng, Z., Shunren, H., & Xiaowei, H. (2009). Review of fault diagnosis in control systems. In *Control and Decision Conference, CCDC '09* (pp. 5324–5329). Chinese.
3. Patton, R. J., Uppal, F. J., & Lopez-Toribio, C. J. (2000). Soft computing approaches to fault diagnosis for dynamic systems-A survey. In *Proceedings of IFAC Symposium on Fault Detection, Supervision and Safety for Technical Processes* (pp. 303–315), Budapet, Hungary
4. Isermann, R. (1984). Process fault detection based on modeling and estimation methods-A survey. *Automatica, 20*, 387–404.
5. Venkatasubramanian, V., Rengaswamy, R., Yin, K., & Kavuri, S. N. (2003a). A review of process fault detection and diagnosis: Part I: Quantitative model-based methods. *Computers and Chemical Engineering, 27*, 293–311.
6. Venkatasubramanian, V., Rengaswamy, R., Kavuri, S. N., & Yin, K. (2003c). A review of process fault detection and diagnosis: Part III: Process history based methods. *Computers and Chemical Engineering, 27*, 327–346.
7. Venkatasubramanian, V., Rengaswamy, R., & Kavuri, S. N. (2003b). A review of process fault detection and diagnosis: Part II: Qualitative models and search strategies. *Computers and Chemical Engineering, 27*, 313–326.
8. Angeli, C., & Chatzinikolaoui, A. (2004). On-line fault detection techniques for technical systems- a survey. *International Journal of Computer Science and Applications, 1*, 12–30.
9. Uraikul, V., Chan, C. W., & Tontiwachwuthikul, P. (2007). Artificial intelligence for monitoring and supervisory control of process systems. *Engineering Applications of Artificial Intelligence, 20*, 115–131.
10. Maurya, M. R., Rengaswamy, R., & Venkatasubramanian, V. (2007). Fault diagnosis using dynamic trend analysis: A review and recent developments. *Engineering Applications of Artificial Intelligence, 20*, 133–146.
11. Abellan-Nebot, J., & Romero, Subir F. (2009). A review of machining monitoring systems based on artificial intelligence process models. *The International Journal of Advanced Manufacturing Technology, 47*, 237–257.
12. Stephanopoulos, G., & Han, C. (1996). Intelligent systems in process engineering: R review. *Computers and Chemical Engineering, 20*, 743–791.
13. Bocaniala, C. D., Palade, V. (2006). Computational intelligence methodologies in fault diagnosis: Review and state of the art. In *Computational intelligence in fault diagnosis* (pp. 1–36).
14. Mozeti, I. (1992). Model-based diagnosis: An overview. In V. Mrk, O. tpnkov, R. Trappl (Eds.), Advanced topics in artificial intelligence. Lecture Notes in Computer Science (Lecture Notes in Artificial Intelligence) (Vol. 617, pp. 419–430). Berlin: Springer.
15. Previdi, F., & Parisini, T. (2006). Model-free actuator fault detection using a spectral estimation approach: the case of the DAMADICS benchmark problem. *Control Engineering Practice, 14*, 635–644.
16. Dorr, R., Kratz, F., Ragot, J., Loisy, F., & Germain, J. L. (1996). Detection, isolation, and identification of sensor faults in nuclear power plants. *IEEE Transactions on Control Systems Technology, 5*, 42–60.

17. Leger, R. P., Garland, W. J., & Poehlman, (1998). Fault detection and diagnosis using statistical control charts and artificial neural networks. *Artificial Intelligence in Engineering, 12*, 35–47.
18. Finn, J., Wagner, J., & Bassily, H. (2010). Monitoring strategies for a combined cycle electric power generator. *Applied Energy, 87*, 2621–2627.
19. Namburu, S. M., Azam, M. S., Jianhui, L., Kihoon, C., & Pattipati, K. R. (2007). Data-driven modeling, fault diagnosis and optimal sensor selection for HVAC chillers. *IEEE Transactions on Automation Science and Engineering, 4*, 469–473.
20. Mourot, G., Bousghiri, S., & Ragot, J. (1993). Pattern recognition for diagnosis of technological systems: A review. In *Systems, Man and Cybernetics, 1993. Conference Proceedings. International Conference on 'Systems Engineering in the Service of Humans* (Vol. 5, pp. 275–281).
21. Ruz-Hernandez, J. A., Sanchez, E. N., & Suarez, D. A. (2005). Neural networks-based scheme for fault diagnosis in fossil electric power plants. In *IJCNN '05. Proceedings. IEEE International Joint Conference on Neural Networks* (Vol. 3, pp. 740–1745).
22. Liang-Yu, M., Jin, M., Xian, G., & Bing-Shu, W. (2005). Fuzzy pattern recognition approach with symptom zoom technology to diagnose a slight and incipient fault in a thermodynamic system. In *Proceedings of International Conference on Machine Learning and Cybernetics* (Vol. 7, pp. 4410–4414).
23. Isermann, R., & Ulieru, M. (1993). Integrated fault detection and diagnosis. In *Systems Engineering in the Service of Humans. Conference Proceedings. International Conference on in Systems, Man and Cybernetics* (Vol. 1, pp. 743–748).
24. Isermann, R. (1998). On fuzzy logic applications for automatic control, supervision, and fault diagnosis. IEEE Trans. *Syst. Man Cybern. Part A Syst. Hum., 28*, 221–235.
25. Clark R. N. (1978). A simplified instrument failure detection scheme. *IEEE Transactions on Aerospace and Electronic Systems, AES-14*, 558–563.
26. Clark, R. N. (1979). The dedicated observer approach to instrument failure detection. In *18th IEEE Conference on Decision and Control including the Symposium on Adaptive Processes* (pp. 237–241).
27. Beard, R. V. (1971). Failure Accommodation in Linear Systems through Self-organization. Ph.D Thesis. Massachusetts Institute of Technology.
28. Seliger, R., & Koppen-Seliger, B. (1995). Robust nonlinear observer-based fault detection for a U-tube steam generator. In *Proceedings of the American Control Conference* (Vol. 2, pp. 1134–1135).
29. Simani, S., Fantuzzi, C., Beghelli, S. (1999). Improved observer for sensor fault diagnosis of a power plant. In *Proceedings of the 7th Mediterranean Conference on Control and Automation Haifa. Israel* (Vol. 1, pp. 826–834).
30. Dassanake, S. K., Balas, G. L., & Bokor, J. (2000). Using unknown input observers to detect and isolate sensor faults in a turbofan engine. In *Digital Avionics Systems Conferences, 2000. Proceedings. DASC* (Vol. 2, pp. 1–7).
31. Astorga-Zaragoza, C. M., Zavala-Ro, A., Alvarado, V. M., Mndez, R. M., & Reyes-Reyes, J. (2007). Performance monitoring of heat exchangers via adaptive observers. *Measurement, 40*, 392–405.
32. Astorga-Zaragoza, C. M., Alvarado-Martnez, V. M., Zavala-Ro, A., Mndez-Ocaa, R. M., & Guerrero-Ramrez, G. V. (2008). Observer-based monitoring of heat exchangers. *ISA Transactions, 47*, 15–24.
33. Odgaard, P. F., & Mataji, B. (2008). Observer-based fault detection and moisture estimating in coal mills. *Control Engineering Practice, 16*, 909–921.
34. Simani, S., Fantuzzi, C., & Beghelli, S. (2000). Diagnosis techniques for sensor faults of industrial processes. *IEEE Transactions on Control Systems Technology, 8*, 848–855.
35. Gertler, J. (1995). Diagnosing parametric faults: From parameter estimation to parity relations. In *Proceedings of the American Control Conference* (Vol. 3, pp. 1615–1620).
36. Spina, P. R. (2000). Reliability in the determination of gas turbine operating state. In *Proceedings of the 39th. IEEE Conference on Decision and Control* (Vol. 3, pp. 2639–2644)

37. Khajali, B., Poshtan, J. (2007). Model-based fault detection using parity equations applied to a petrochemical system. In *ICIAS International Conference on Intelligent and Advanced Systems* (pp. 1141–1146).
38. Fagarasan I., & St. Iliescu S. (2008) Parity equations for fault detection and isolation. In *IEEE International Conference on Automation, Quality and Testing, Robotics. AQTR 2008* (pp. 99–103).
39. Bloch, G., Ouladsine, M., & Thomas, P. (1995). On-line fault diagnosis of dynamic systems via robust parameter estimation. *Control Engineering Practice, 3*, 1709–1717.
40. Moseler, O., & Isermann R. (1998). Model-based fault detection for a brushless DC motor using parameter estimation. In *Industrial Electronics Society, IECON '98. Proceedings of the 24th Annual Conference of the IEEE* (Vol. 4, pp. 1956–1960).
41. De La Fuente, M. J., & Vega, P. (1995). A neural networks based approach for fault detection and diagnosis: Application to a real process. In *Proceedings of the 4th IEEE Conference on Control Applications* (pp. 188–193).
42. Balle, P., & Isermann, R. (1998). Fault detection and isolation for nonlinear processes based on local linear fuzzy models and parameter estimation. In *Proceedings of the American Control Conference* (Vol. 3, pp. 1605–1609).
43. Abidin, M. S. Z., Yusof, R., Kahlid, M., & Amin, S. M. (2002). Application of a model-based fault detection and diagnosis using parameter estimation and fuzzy inference to a DC-servomotor. In *Proceedings of the 2002 IEEE International Symposium on Intelligent Control* (pp. 783–788).
44. Weyer, E., Szederknyi, G., & Hangos, K. (2000). Grey box fault detection of heat exchangers. *Control Engineering Practice, 8*, 121–131.
45. Kumar, S., Sinha, S., Kojima, T., & Yoshida, H. (2001). Development of parameter based fault detection and diagnosis technique for energy efficient building management system. *Energy Conversion and Management, 42*, 833–854.
46. Yoshida, H., Kumar, S., & Morita, Y. (2001). Online fault detection and diagnosis in VAV air handling unit by RARX modeling. *Energy and Buildings, 33*, 391–401.
47. Tylee, J. (1983). On-line failure detection in nuclear power plant instrumentation. *IEEE Transactions on Automatic Control, 28*, 406–415.
48. Usoro, P. B., Schick, I. C., & Negahdaripour, S. (1985). HVAC system fault detection and diagnosis. In *American Control Conference* (pp. 606–612).
49. Li, R. (1991). Fault detection and diagnosis: Extended Kalman filters and neural networks Thesis. United States – Delaware: University of Delaware USA. UMI Order No. GAX92-06377.
50. Hines, J. W. (1994). A hybrid approach to fault detection and isolation: Merging analytical redundancy and neural network techniques. Thesis. United States – Ohio: The Ohio State University. https://elibrary.ru/item.asp?id=5633812
51. Simani, S., & Fantuzzi, C. (2000). Fault diagnosis in power plant using neural networks. *Information Sciences, 127*, 125–136.
52. Yang, Q. (2004). Model-based and data driven fault diagnosis methods with applications to process monitoring. Thesis. United States – Ohio: Case Western Reserve University. https://etd.ohiolink.edu/rws_etd/document/get/case1080246972/inline
53. Zhang, S., Toshiyuki, A., Shoji, H. (2004). Gas leakage detection system using Kalman filter. In *7th International Conference on Signal Processing, Proceedings. ICSP '04* (Vol. 3, pp. 2533–2536).
54. Ryerson, C. (2006). *A fault detection scheme for modeled and unmodeled faults in a simple hydraulic actuator system using an extended Kalman filter. Thesis.* United States - Missouri: University of Missouri - Columbia.
55. Jackson, J. E. (1991). *A User's Guide to Principal Components* (Vol. 587). USA: Wiley Inc.
56. Tien, D. X., Lim, K. W., Jun, L. (2004). Comparative study of PCA approaches in process monitoring and fault detection. In *30th Annual Conference of IEEE Industrial Electronics Society, IECON* (Vol. 3, pp. 2594–2599).

57. Jang, K. J. (1999). Process monitoring and fault classification for an air handling unit. Thesis. United States – Iowa: Iowa State University. http://lib.dr.iastate.edu/cgi/viewcontent.cgi?article=13571&context=rtd

58. Wang, S., & Xiao, F. (2004). Detection and diagnosis of AHU sensor faults using principal component analysis method. *Energy Conversion and Management, 45*, 2667–2686.

59. Cui, J. (2005). *A robust fault detection and diagnosis strategy for centrifugal chillers.* Thesis. Hong Kong: Hong Kong Polytechnic University (Hong Kong).

60. Wang, S., & Cui, J. (2005). Sensor-fault detection, diagnosis and estimation for centrifugal chiller systems using principal-component analysis method. *Applied Energy, 82*, 197–213.

61. Xu, X., Xiao, F., & Wang, S. (2008). Enhanced chiller sensor fault detection, diagnosis and estimation using wavelet analysis and principal component analysis methods. *Applied Thermal Engineering, 28*, 226–237.

62. Ke, Z., & Upadhyaya, B. R. (2006). Model based approach for fault detection and isolation of helical coil steam generator systems using principal component analysis. *IEEE Transactions on Nuclear Science, 53*, 2343–2352.

63. Upadhyaya, B. R., Zhao, K., & Lu, B. (2003). Fault monitoring of nuclear power plant sensors and field devices. *Progress in Nuclear Energy, 43*, 337–342.

64. Donat, W., Choi, K., An, W., Singh, S., & Pattipati, K. (2008). Data visualization, data reduction and classifier fusion for intelligent fault diagnosis in gas turbine engines. *Journal of Engineering for Gas Turbines and Power, 130*, 1–8.

65. Mina, J., Verde, C., Sanchez-Parra, M., & Ortega, F. (2008). Fault isolation with principal components structured models for a gas turbine. In *American Control Conference* (pp. 4268–4273).

66. Li, S. (2009). A model-based fault detection and diagnostic methodology for secondary HVAC systems. Thesis. United States – Pennsylvania: Drexel University https://idea.library.drexel.edu/islandora/object/idea

67. Qin, J. (2006). A fault detection and diagnosis strategy for VAV air distribution system. HKIE transactions, Hong Kong: Hong Kong Polytechnic University (Hong Kong) (Vol. 13, pp. 36–43). http://ira.lib.polyu.edu.hk/handle/10397/25236

68. Shaoyuan, Z., Jianming, Z., & Shuqing, W. (2004). Fault diagnosis in industrial processes using principal component analysis and hidden Markov model. In *Proceedings of the American Control Conference* (Vol. 6, pp. 5680–5685).

69. Hadad, K., Mortazavi, M., Mastali, M., & Safavi, A. A. (2008). Enhanced neural network based fault detection of a VVER nuclear power plant with the aid of principal component analysis. *IEEE Transactions on Nuclear Science, 55*, 3611–3619.

70. Maghsooloo, A., Khosravi, A., Anvar, H. S., Barzamini, R. (2008). ANFIS and PCA capability assessment for fault detection in unknown nonlinear systems. In *3rd International Symposium on Communications, Control and Signal Processing, ISCCSP*. 47-b-52.

71. Hsu, C. C., Chen, M. C., & Chen, L. S. (2010). A novel process monitoring approach with dynamic independent component analysis. *Control Engineering Practice, 18*, 242–253.

72. Cybennko, G. (1989). Approximation by superpositions of a sigmoidal function. *Mathematics of Control, Signals, and Systems, 2*, 303–314.

73. Wenli, Y., Lee, K. Y., Junker, S. T., & Ghezel-Ayagh, H. (2008). Fault diagnosis and accommodation system with a hybrid model for fuel cell power plant. *Power and energy society general meeting - conversion and delivery of electrical energy in the 21st century* (pp. 1–8). IEEE: Pittsburgh.

74. Szczepaniak, P. S. (1994). Application of neural networks for fault diagnosis in a power plant. In *Second International Conference on Intelligent Systems Engineering* (pp. 292–297).

75. Rakhshani, E., Sariri, I., & Rouzbehi, K. (2009). Application of data mining on fault detection ad prediction in Boiler of power plant using artificial neural network. In *International Conference on Power Engineering, Energy and Electrical Drives, POWERENG '09* (pp. 473–478).

76. Sanchez, E. N., Suarez, D. A., & Ruz, J. A. (2004). Fault detection in fossil electric power plant via neural networks. In *Automation Congress, Proceedings. World* (pp. 213–218).

77. Karlsson, C., Arriagada, J., & Genrup, M. (2008). Detection and interactive isolation of faults in steam turbines to support maintenance decisions. *Simulation Modelling Practice and Theory, 16*, 1689–1703.

78. Hwang, B. C. (1993). *Fault detection and diagnosis of a nuclear power plant using artificial neural networks*. Thesis. Canada: Simon Fraser University, Canada.

79. Guglielmi, G., Parisini, T., & Rossi, G. (1995). Keynote paper: Fault diagnosis and neural networks: A power plant application. *Control Engineering Practice, 3*, 601–620.

80. Furukawa, H., Ueda, T., & Kitamura, M. (1996). A systematic method for rational definition of plant diagnostic symptoms by self-organizing neural networks. *Neurocomputing, 13*, 171–183.

81. Keyvan, S. (2001). Traditional signal pattern recognition versus artificial neural networks for nuclear plant diagnostics. *Progress in Nuclear Energy, 39*, 1–29.

82. Ogaji, S. O. T., & Singh, R. (2003). Advanced engine diagnostics using artificial neural networks. *Applied Soft Computing, 3*, 259–271.

83. Koppen-Seliger, B., Frank, P. M. (1995). Fault detection and isolation in technical processes with neural networks. In *Proceedings of the 34th IEEE Conference on Decision and Control* (Vol. 3, pp. 2414–2419)

84. Hoskins, J. C., Kaliyur, K. M., & Himmelblau, D. M. (1990). Incipient fault detection and diagnosis using artificial neural networks. *IJCNN International Joint Conference on Neural Networks,* (Vol. 1, pp. 81–86).

85. Pin-Hsuan, W., Jen-Pin, Y., Fang-Tsung, L., Huang-Chu, H., Ker-Wei, Y., Rey-Chue, H. (2008). An intelligent fault diagnostics for turbine generator by modified neural model. In *3rd International Conference on Innovative Computing Information and Control, ICICIC '08* (pp. 296–296).

86. Chun-ling, X., Jen-Yuan, C., Xiao-cheng, S., & Jing-min, D. (2008). Fault diagnosis of nuclear power plant based on genetic-rbf neural network. In *15th International Conference on Mechatronics and Machine Vision in Practice, M2VIP* (pp. 334–339).

87. Embrechts, M. J., & Benedek, S. (1997). Identification of nuclear power plant transients with neural networks. In *IEEE International Conference on Systems, Man, and Cybernetics, Computational Cybernetics and Simulation,* (Vol. 1, pp. 912–916).

88. Ferreira, P. B. (1999). *Incipient fault detection and isolation of sensors and field devices*. Thesis, United States - Tennessee: The University of Tennessee.

89. Lu, B., & Upadhyaya, B. R. (2005). Monitoring and fault diagnosis of the steam generator system of a nuclear power plant using data-driven modeling and residual space analysis. *Annals of Nuclear Energy, 32*, 897–912.

90. Witczak, M., Korbicz, J., Mrugalski, M., & Patton, R. J. (2006). A GMDH neural network-based approach to robust fault diagnosis: Application to the DAMADICS benchmark problem. *Control Engineering Practice, 14*, 671–683.

91. Li, F., Upadhyaya, B. R., & Coffey, L. A. (2009). Model-based monitoring and fault diagnosis of fossil power plant process units using group method of data handling. *ISA Transactions, 48*, 213–219.

92. Sreedhar, R., Fernandez, B., & Masada, G. Y. (1992). A neural network based adaptive fault detection scheme. In *Proceedings of the American Control Conference* (Vol. 5, pp. 3259–3263).

93. Weerasinghe, M., Gomm, J. B., & Williams, D. (1998). Neural networks for fault diagnosis of a nuclear fuel processing plant at different operating points. *Control Engineering Practice, 6*, 281–289.

94. Guo, Z., & Uhrig, R. E. (1992). Using genetic algorithms to select inputs for neural networks. In *International Workshop on Combinations of Genetic Algorithms and Neural Networks. COGANN-92* (pp. 223–234).

95. Simani, S., Fantuzzi, C., & Spina, R. P. (1998). Application of a neural network in gas turbine control sensor fault detection. In *Proceedings of the 1998 IEEE International Conference on Control Applications* (Vol. 1, pp. 182–186).

96. Simani, S. (2005). Identification and fault diagnosis of a simulated model of an industrial gas turbine. *IEEE Transactions on Industrial Informatics, 1*, 202–216.
97. Fantoni, P. F., & Mazzola, A. (1996). A pattern recognition-artificial neural networks based model for signal validation in nuclear power plants. *Annals of Nuclear Energy, 23*, 1069–1076.
98. Nabeshima, K., Suzudo, T., Seker, S., Ayaz, E., Barutcu, B., Trkcan, E., et al. (2003). On-line neuro-expert monitoring system for Borssele nuclear power plant. *Progress in Nuclear Energy, 43*, 397–404.
99. Kramer, M. A. (1991). Nonlinear principal componenet analysis using autoassociative neural networks. *AIChE Journal, 37*, 233–243.
100. Ogaji, S., Sampath, S., Singh, R., & Probert, D. (2002). Novel approach for improving power-plant availability using advanced engine diagnostics. *Applied Energy, 72*, 389–407.
101. Lu, P. J., Zhang, M. C., Hsu, T. C., & Zhang, J. (2001). An evaluation of engine faults diagnostics using artificial neural networks. *Journal of Engineering for Gas Turbines and Power, 123*, 340–346.
102. Bourassa, M. A. J. (1999). Autoassociative neural networks with an application to fault diagnosis of a gas turbine engine, (2000), 0724–0724 (p. 248). Canada: Royal Military College of Canada (Canada).
103. Diao, Y., & Passino, K. M. (2004). Fault diagnosis for a turbine engine. *Control Engineering Practice, 12*, 1151–1165.
104. Ogaji, S. O. T., Marinai, L., Sampath, S., Singh, R., & Prober, S. D. (2005). Gas-turbine fault diagnostics: A fuzzy-logic approach. *Applied Energy, 82*, 81–89.
105. Koppen-Seliger, B., Kiupel, N., Schulte Kellinghaus, H., Frank, P. M. (1995). A fault diagnosis concept for a high-pressure-preheater line. In *Proceedings of the 34th IEEE Conference on Decision and Control* (Vol. 3, pp. 2383–2388).
106. Zhao, K., & Upadhyaya, B. R. (2005). Adaptive fuzzy inference causal graph approach to fault detection and isolation of field devices in nuclear power plants. *Progress in Nuclear Energy, 46*, 226–240.
107. Heo J. S., & Lee K. Y. (2006). A multi-agent system-based intelligent identification system for power plant control and fault-diagnosis. In *Power Engineering Society General Meeting, 2006. IEEE 2006 Jun* (pp. 6). New York: IEEE.
108. Arranz, A., Cruz, A., Sanz-Bobi, M. A., Ruz, P., & Coutio, J. (2008). DADICC: Intelligent system for anomaly detection in a combined cycle gas turbine plant. *Expert Systems with Applications, 34*, 2267–2277.
109. Ogaji, S. O. T., & Singh, R. (2002). Advanced engine diagnostics using artificial neural networks. In *2002 IEEE International Conference on Artificial Intelligence Systems, (ICAIS 2002)* (pp. 236–241).
110. Power, Y., & Bahri, P. A. (2004). A two-step supervisory fault diagnosis framework. *Computers and Chemical Engineering, 28*, 2131–2140.
111. Palade, V., Patton, R. J., Uppal, F. J., Quevedo, J., Daley, S. (2002). Fault Diagnosis of an Industrial Gas Turbine Using Neuro-Fuzzy Methods. In *Proceedings of the 15th IFAC World Congress Barcelona* (pp. 2477–2482).
112. Ayoubi, M., & Isermann, R. (1997). Neuro-fuzzy systems for diagnosis. *Fuzzy Sets and Systems, 89*, 289–307.
113. Jang, J. S. R. (1993). ANFIS: adaptive-network-based fuzzy inference system. *IEEE Transactions on Systems, Man and Cybernetics, 23*, 665–685.
114. Negoita M. G., Neagu D., & Palade V. (2005). Neuro-Fuzzy Integration in Hybrid Intelligent Systems. In *Computational intelligence* (pp. 25–39).
115. Lin, C. T., & Lee, C. S. G. (1991). Neural-network-based fuzzy logic control and decision system. *IEEE Transactions on Computers, 40*, 1320–1336.
116. Nelles, O. (2001). *Nonlinear system identification*. Berlin: Springer.
117. Korbicz, J., Koscielny, J. M., & Kowalczuk, Z. (2004). *Fault diagnosis: Models, artificial intelligence, applications*. Berlin: Spriinger.
118. Zhang, J., & Morris, J. (1996). Process modelling and fault diagnosis using fuzzy neural networks. *Fuzzy Sets and Systems, 79*, 127–140.

119. Zio, E., & Gola, G. (2006). Neuro-fuzzy pattern classification for fault diagnosis in nuclear components. *Annals of Nuclear Energy, 33*, 415–426.

120. Uppal, F. J., Patton, R. J., & Witczak, M. (2006). A neuro-fuzzy multiple-model observer approach to robust fault diagnosis based on the DAMADICS benchmark problem. *Control Engineering Practice, 14*, 699–717.

121. Bartys, M., Patton, R., Syfert, M., De Las Heras, S., & Quevedo, J. (2006). Introduction to the DAMADICS actuator FDI benchmark study. *Control Engineering Practice, 14*, 577–596.

122. Korbicz, J., & Kowal, M. (2007). Neuro-fuzzy networks and their application to fault detection of dynamical systems. *Engineering Applications of Artificial Intelligence., 20*, 609–617.

123. Tan, S. C., Rao, M. V. C., & Lim, C. P. (2008). Fuzzy ARTMAP dynamic decay adjustment: An improved fuzzy ARTMAP model with a conflict resolving facility. *Applied Soft Computing, 8*, 543–554.

124. Chen K., Lim C., & Lai W. (2006) Fault Diagnosis in a Power Generation Plant Using a Neural Fuzzy System with Rule Extraction. In *Computational intelligence in fault diagnosis* (pp. 287–304).

125. Razavi-Far, R., Davilu, H., Palade, V., & Lucas, C. (2009). Model-based fault detection and isolation of a steam generator using neuro-fuzzy networks. *Neurocomputing, 72*, 2939–2951.

126. Evsukoff, A., & Gentil, S. (2005). Recurrent neuro-fuzzy system for fault detection and isolation in nuclear reactors. *Advanced Engineering Informatics, 19*, 55–66.

127. Adamson, M. S. (1990). Development of a diagnostic expert system using fault tree analysis information. MEng Thesis, Canada: Royal Military College of Canada (Canada).

128. Chen, L. W. (1992). *A hierarchical decision process for fault administration. Thesis*. United States - Maryland: University of Maryland College Park.

129. McDonald, J. R., Stewart, R. W., Gemmell, B. D., Georgin, E. J. M., & Weir, B. J. (1991). Advanced condition monitoring and fault diagnosis of turbo-alternators using an embedded expert system. In *IEE Colloquium on: Advanced Condition Monitoring Systems for Power Generation* (pp. 1–4).

130. Berrios R., Nunez F., Cipriano A., & Paredes R. (2008). Expert fault detection and diagnosis for the refrigeration process of a hydraulic power plant. In *27th Chinese Control Conference, CCC 2008* (pp. 122–126)

131. Mihara, K., Aono, Y., Ohkawa, T., Komoda, N., & Miyasaka, F. (1994). Stochastic qualitative reasoning and its application to diagnosis of air conditioning system. In *20th International Conference on Industrial Electronics, Control and Instrumentation, IECON '94* (Vol. 2, pp. 1401–1406).

132. Yong-Guang, M., Jian-Qiang, G., Liang-Yu, M., Qin, Y., & Peng, T. (2006). Study on fault diagnosis based on the qualitative/quantitative model of SDG and genetic algorithm. In *International Conference on Machine Learning and Cybernetics* (pp. 2053–2058).

133. Ji, Z., Wen-liang, C., Bing-shu, W., & Ning, C. (2005). Fault location algorithm based on the qualitative and quantitative knowledge of signed directed graph. In *IEEE International Conference on Industrial Technology, 2005. ICIT 2005* (pp. 1231–1234).

134. Conroy, G. V., Black, W. J., & O'Hare, G. M. P. (1989). Fault diagnosis and on condition maintenance of a combined heat and power unit. In *IEE Colloquium on: Expert Systems for Fault Diagnosis in Engineering Applications* (pp. 1–5).

135. Sztipanovits, J., Karsai, G., Padalkar, S., Biegl, C., Miyasaka, N., & Okuda, K. (1990). Intelligent monitoring and diagnostics for plant automation. In *Proceedings, IEEE International Conference on Robotics and Automation* (Vol. 2, pp. 1390–1395).

136. Padalkar, S., Karsai, G., Sztipanovits, J., Okuda, K., & Miyasaka, N. (1991). Real-time fault diagnostics with multiple aspect models. In *Proceedings, IEEE International Conference on Robotics and Automation* (Vol. 1, pp. 803–808).

137. Kumamaru, T., Utsunomiya, T., Yamada, Y., Iwasaki, Y., Shoda, I., & Obayashi, M. (1991). A fault diagnosis system for district heating and cooling facilities. In *International Conference on Industrial Electronics, Control and Instrumentation, Proceedings. IECON '91* (Vol. 1, pp. 131–136).

138. Perryman, R., & Perrott, S. N. (1994). Condition monitoring and predictive analysis of combined heat and power systems. In *International Conference on Life Management of Power Plants* (pp. 40–47).

139. Perryman, R. (1995). Condition monitoring of combined heat and power systems. In *IEE Colloquium on: Condition Monitoring of Electrical Machines* (pp. 1–4).

140. Renders, J. M., Goosens, A., De Viron, F., & De Vlaminck, M. (1995). A prototype neural network to perform early warning in nuclear power plant. *Fuzzy Sets and Systems, 74*, 139–151.

141. Batanov, D. N., & Cheng, Z. (1995). An object-oriented expert system for fault diagnosis in the ethylene distillation process. *Computers in Industry, 27*, 237–249.

142. Bonarini, A., & Sassaroli, P. (1997). Opportunistic multimodel diagnosis with imperfect models. *Information Sciences, 103*, 161–185.

143. Muoz, A., & Sanz-Bobi, M. A. (1998). An incipient fault detection system based on the probabilistic radial basis function network: Application to the diagnosis of the condenser of a coal power plant. *Neurocomputing, 23*, 177–194.

144. Abu-el-zeet, Z. H., & Patel, V. C. (2006). Power plant condition monitoring using novelity detection. In *International Conference on Systems Engineering. Systems Engineering, Coventry University (ICSE2006)* (pp. 9–14).

145. Biagetti, T., & Sciubba, E. (2004). Automatic diagnostics and prognostics of energy conversion processes via knowledge-based systems. *Energy, 29*, 2553–2572.

146. Sciubba, E. (2004). Hybrid semi-quantitative monitoring and diagnostics of energy conversion process. *International Journal of Thermodynamics, 7*, 95–106.

147. Valero, A., Correas, L., Lazzaretto, A., Rangel, V., Reini, M., Taccani, R., et al. (2004). Thermoeconomic philosophy applied to the operating analysis and diagnosis of energy utility systems. *International Journal of Thermodynamics, 7*, 33–39.

148. Lazzaretto, A., & Toffolo, A. (2006). A critical review of the thermoeconomic diagnosis methodologies for the location of causes of malfunctions in energy systems. *ASME Journal of Energy Resources Technology, 128*, 335–342.

149. Thomson, M., Twigg, P. M., Majeed, B. A., & Ruck, N. (2000). Statistical process control based fault detection of CHP units. *Control Engineering Practice, 8*, 13–20.

150. Flynn, D., Ritchie, J., & Cregan, M. (2005). Data mining techniques applied to power plant performance monitoring. *IFAC Proceedings Volumes, 38*(1), 369–74.

151. Niu, Z., Liu, J. Z., Niu, Y.G., & Pan, Y.S. (2005). A reformative PCA-based fault detection method suitable for power plant process. In *Proceedings of 2005 International Conference on Machine Learning and Cybernetics, Aug 18 2005* (Vol. 4, pp. 2133–2138). Guangzhou: IEEE.

152. Odgaard, P. F., & Mataji, B. (2006). Fault detection in coal mills used in power plants. *IFAC Proceedings Volumes. 2006 Dec 31, 39*(7), 177–182.

153. Hines, J. W., Don, W. M., Wesley, J., Hines, D., Miller, W., & Brian K. H. (2007) Fault detection and isolation: A hybrid approach. In *Proceedings of the 1995 American Nuclear Society Annual Meeting and Embedded Topical Meeting on Computer-Based Human Support Systems: Technology, Methods and Future, Philadelphia.* https://pdfs.semanticscholar.org/b5a0/89eccb2cb05625413d2c9f61eafef9f763fb.pdf

154. Korodi, A., & Dragomir, T. L. (2007). Mobile fault detection and diagnosis module for automatic systems. In *Mediterranean Conference on Control and Automation, 2007. MED'07* (pp. 1–6). Greece: IEEE.

155. Fast, M., & Palm, T. (2010). Application of artificial neural networks to the condition monitoring and diagnosis of a combined heat and power plant. *Energy, 35*, 1114–1120.

Chapter 3
Model Identification Using Neuro-Fuzzy Approach

3.1 Introduction

This chapter contains the discussion on fundamental concepts related to nonlinear model identification. First, linear in parameter model identification techniques are presented. This covers static and dynamic systems. Following that, the idea of developing nonlinear models in the framework of Orhonormal Basis Functions (OBF) is described. In Sect. 3.3, basic theory of neural networks and fuzzy systems are elaborated. In the state of the art designs, one of them is constructed in the structure of the other allowing the development of a transparent model that can be trained with relatively minimal effort. Section 3.4 is dedicated to the discussion of nonlinear system identification using combined version of neural networks and fuzzy systems. Last section of the chapter deals with three different model training algorithms Least squares based, back-propagation and particle swarm optimization.

3.2 Model Identification

The need to understand a system necessitates an experiment on the system or replacing the system by an abstract model. The latter is advantages as it provides flexibility in trying to test the system outside acceptable regions too. If a model is developed applying known laws of nature conservation of energy, conservation of mass, conservation of momentum, state equations, and Fourier Heat Transfer, then the model developed is said to be a first principle model. If the model is developed based on input output data taken either from actual experiment or from simulation of high fidelity first principle models, the technique is called model identification.

Identification methods are helpful in the following cases:

- When first principle model coefficients are difficult to determine caused by difficulties in understanding the system very well,

© Springer International Publishing AG 2018
T.A. Lemma, *A Hybrid Approach for Power Plant Fault Diagnostics*,
Studies in Computational Intelligence 743,
https://doi.org/10.1007/978-3-319-71871-2_3

- When the design information are partially known for proprietary reasons, and
- When there is a need to replace the high fiedlty model by a surrogate or meta model for optimization reasons.

A general linear dynamic continuous system can be described by the following differential equation:

$$a_n \frac{d^n y}{dt^n} + a_{n-1} \frac{d^{n-1} y}{dt^{n-1}} + \cdots + a_1 \frac{dy}{dt} + a_o y = b_m \frac{d^m u}{dt^m} + \cdots + b_1 \frac{du}{dt} + b_o u$$

where, $u \in R$ is the input, $y \in R$ is the output, and $a_i (i = 0, 1, 2, \ldots, n)$ and $b_j (j = 0, 1, 2, \ldots, m)$ are the model parameters. Assuming that T represents the sampling time, the equivalent form in discrete time for the differential equation model is

$$y(t + nT) + a_1 y(t + (n - T)) + \cdots a_n y(t) = b_o u(t + nT) + \cdots b_m u(t)$$

For a linear system the coefficients of the system equations do not depend on, $u \in R, y \in R$ and their derivatives. For a time-invariant system, they dont rely on time either. If any one of the stated conditions is included, then the system is said to be nonlinear. The classification between linear and nonlinear models is based on the principle of superposition. In fact cogeneration and cooling plants are nonlinear and piece-wise linear models are used because of their simplicity. If superposition does not hold, then it is said to be nonlinear.

The structure of a process may provide partial information about the system. In that case the identification method is mostly focused on estimating the unknown model parameters. If such a model is formed, it is known as tailor-made model (Grey-Box Model). Models with no physical interpretation (Black-Box Model) in their parameters can also be developed if the system is too complex to sort out the interconnections of component parts applying first principles. The black-box approach is an attractive one in capturing different measurement trends for the purpose of fault detection and diagnosis. Following Ljung [1], and Sderstrm and Stoica [2], for a Linear Time Invariant (LTI) system, the black-box model is defined as

$$y(t) = G(q^{-1}, \theta)u(t) + H(q^{-1}, \theta)e(t) \tag{3.1}$$

where, $\theta \in R^{n_\theta}$ is vector of model parameters; q^{-1} is the shift or delay operator; $G(q^{-1}, \theta)$ is the input transfer function; $H(q^{-1}, \theta)$ is the noise transfer function; $e(t) \in R$ is measurement noise. $e(t)$ is often assumed independently and identically distributed (iid) with zero mean and variance σ^2 or bounded in a certain interval. The term $G(q^{-1}, \theta)u(t)$ is often referred as deterministic part of the model while $H(q^{-1}, \theta)e(t)$ is know as stochastic part of the model.

In (3.1), the transfer functions are often defined as

$$G(q^{-1}, \theta) = \frac{B(q)}{A(q)} \tag{3.2}$$

$$H(q^{-1}, \theta) = \frac{C(q)}{D(q)} \tag{3.3}$$

where,

$$A(q) = 1 + a_1 q^{-1} + a_2 q^{-2} + \cdots + a_{na} q^{-na},$$

$$B(q) = 1 + b_1 q^{-1} + b_2 q^{-2} + \cdots + b_{nb} q^{-nb},$$

$$C(q) = 1 + c_1 q^{-1} + c_2 q^{-2} + \cdots + c_{nc} q^{-nc},$$

$$(Dq) = 1 + d_1 q^{-1} + d_2 q^{-2} + \cdots + d_{nd} q^{-nd},$$

Five forms of black-box model structures can be established from (3.1). If rational functions $G(q^{-1}, \theta) = B(q)/F(q)$ and $H(q^{-1}, \theta) = C(q)/D(q)$ are assumed, the Box-Jenkins (BJ) model evolves. If $H(q^{-1}, \theta) = C(q)/D(q)$ is set equal to 1, as such the characteristics of the error signal is not of concern, then the resulting model is known as Output Error (OE) model. In some cases the functions $F(q)$ and $D(q)$ each are set equal to $A(q)$. In that case the original equation takes a different form, which is known as AutoRegression Moving Average eXogenous input (ARMAX) model. If the characteristic function of the error term $e(t)$, i.e. $C(q)$, is set equal to zero, then it will be reduced to the model called AutoRegression eXtra (ARX) input model. The last and simplified model is the Finite Impulse Response (FIR) where $G(q^{-1}, \theta) = B(q)$ and $H(q^{-1}, \theta) = 1$. Figure 3.1 presents block diagram representation of each model.

BJ, ARMAX and OE models involve autoregressive part. Assuming that the transfer function for the stochastic part of the black-box model is invertible, a reduced model suitable for prediction can be deduced from (3.1). This model is given by (3.4). Note that the first and the second terms in the right hand side of the equation carry past values of $y(t)$ and $u(t)$, respectively. The only term unknown in the right side of the equation is $e(t)$. By eliminating $e(t)$ from (3.2), a one step ahead prediction on $y(t)$ can be estimated.

$$y(t) = (1 - H(q^{-1}, \theta)^{-1}) y(t) + H(q^{-1}, \theta)^{-1} G(q, \theta) U(t) + e(t) \tag{3.4}$$

Equations (3.1) and (3.4) work for LTI systems. The counterpart for a nonlinear system is formulated assuming a nonlinear relationship between the system past inputs, past outputs, additive disturbances and current system outputs. That is,

$$y(t) = f(x(t), \theta) + \varepsilon(t) \tag{3.5}$$

where, $x(t) = [-y(t-1) \cdots - y(t-n_y), u(t), u(t-1), \ldots, u(t-n_u)]$, and $f(.)$ is the nonlinear mapping from R^{n-u+n_y} to R. Now, depending upon the form of $x(t)$, such models named as Nonlinear Box-Jankins (NBJ), Nonlinear AutoRegressive Moving Average eXogenous input (NARMAX), Nonlinear Output Error (NOE) and Nonlinear Finite Impulse Response (NFIR) can be developed. To realized (3.5),

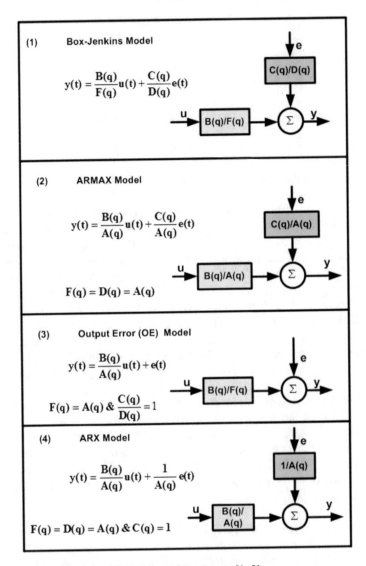

Fig. 3.1 Linear time invariant Black-Box model structures [1, 2]

one may rely on polynomials, Volterra models, piece-wise linear approximation techniques, neural networks or fuzzy models. In all the cases, nonetheless, the model orders n_u and n_y, have to be decided prior to estimating the model parameters. Besides, to ensure that the required accuracy is achieved the model order has to be selected large, which is often the case for FIR and NFIR models. Recent studies, however, suggest that a parsimonious model can be developed if Orthonormal Basis Functions (OBF) are used in the deterministic part of the models for LTI systems and for creating a virtual memory for the nonlinear methods. This approach is also found powerful to avoid any autoregressive part while at the same time providing model accuracy as close as its counterpart.

3.2.1 Orthonormal Basis Functions

OBFs are typical of the specialised basis functions recently getting more attention in model identification and control. The inclusion of these functions creates a room to incorporate prior knowledge about the system dynamics. Two of the OBFs mostly used for model identification and control are Laguerre Function and Kaurtz Function [3]. Starting with the basic definition, two functions are said to be orthonormal if they are orthogonal to each other and if each demonstrates Euclidian norm of unity; that is $\|\phi_i(q-1)\| = \|\phi_k(q-1)\| = 1$. In the frequency domain, the orthogonality condition between two functions $\phi_i(j\omega)$ and $\phi_k(j\omega)$ are satisfied if and only if

$$\langle \phi_i, \phi_k \rangle = \frac{1}{2\pi} \int_{-\infty}^{\infty} \phi_i(j\omega)\phi_k(j\omega)d\omega = \begin{cases} 0 & \text{if } i \neq k \\ 1 & \text{if } i = k \end{cases} \qquad (3.6)$$

where, $\langle ., . \rangle$ is inner product between functions $\phi_i(j\omega)$ and $\phi_k(j\omega)$, and q^{-1} is the shift operator. The basis functions considered for discussion are given in Fig. 3.2. The first kind called Laguerre Basis Function (LBF) [4], Fig. 3.2a, is featured by a single pole ξ, with the condition that $\|\xi\| < 1$. LBF based model is appropriate for a system with a dominant first order model. The impulse and step response of the first three LBF for an assumed pole of $\xi = 0.8$ is shown in Fig. 3.3.

Meixner Basis Functions (MBF) [5], Fig. 3.2b: These functions in z-transform are obtained from transformation of the discrete LBF by appropriate matrix. Similar to LBF, a single pole with the criteria of $\|\xi\| < 1$ is used to generate the basis. However, the additional term "β" that stands for order of generalization makes it accommodate delays in the system. It is shown in [6] that LBF and pulse functions are special cases of MBF; for $\beta = 0$, MBF reduces to LBF. With respect to LBF, MBF is less applied in model identification and control.

Kurtz Basis Function (KBF) [7], Fig. 3.2c, is the third type of OBF applied for model identification. Unlike the LBF, it involves two conjugate poles. For a system having dominant second order dynamics, models based on these functions have been found suitable. The impulse and step response of the first three KBFs for assumed poles of $\xi_{1,2} = 0.6 \pm j0.6$ are depicted in Fig. 3.4. Different from LBF responses, the KBF responses are featured by more oscillation. Laguerre and Kautz functions are special cases of generalized orthonormal basis functions.

Generalized Orthonormal Basis Functions (GOBF), Fig. 3.2d: In this case a mix of different poles is used to define the bases. In 1995, Heuberger et al. [8] demonstrated the possibility of generating OBFs with repeated and fixed poles by assuming a minimum balanced State Space (SS) realization of inner functions with McMillan

(a)

Laguerre Basis Function (LBF):

$$\phi_i(q) = \sqrt{1-\xi^2}\,\frac{(1-\xi q)^{i-1}}{(q-\xi)^i},\ \|\xi\| < 1$$

(b)

Meixner Basis Function (MBF):

$$M_k^{(\beta)}(q) = \left(1-\xi^2\right)^{\beta+1/2}\left(\frac{q}{q-\xi}\right)^{\beta+1}\sum_{j=0}^{k}L_{k+1,j+1}^{(\beta)}\left(\frac{1-\xi q}{q-\xi}\right)^j,\ \|\xi\| < 1$$

(c)

Kurtz Basis Function (KBF):

$$\phi_{2i-1}(q) = \frac{\sqrt{(1-a^2)(1-b^2)}}{q^2+a(b-1)q-b}\left[\frac{-bq^2+a(b-1)q+1}{q^2+a(b-1)q-b}\right]^{i-1}$$

$$\phi_{2i}(q) = \frac{\sqrt{(1-b^2)}(q-a)}{q^2+a(b-1)q-b}\left[\frac{-bq^2+a(b-1)q+1}{q^2+a(b-1)q-b}\right]^{i-1}$$

$$-1 < a < 1 \ \text{ and } \ -1 < b < 1$$

(d)

Generalized Orthonormal Baisis Function (GOBF)

$$\phi_i(q,\xi) = \frac{\sqrt{1-\|\xi_i\|^2}}{(q-\xi_i)}\prod_{j-1}^{i}\frac{(1-\xi_j^* q)}{(q-\xi_j)},\ j=1,2,\ldots,n_b \qquad \xi_i = \xi_R + j\xi_I$$

(e)

Markov Orthonormal Basis Function

$$\phi_i(q) = q^i,\ \text{ for } \ i=1,2,\ldots,d$$

$$\phi_{i+d}(q,\xi) = \frac{\sqrt{1-\|\xi_i\|^2}}{(q-\xi_i)}\prod_{j-1}^{i}\frac{(1-\xi_j^* q)}{(q-\xi_j)}q^{-d},\ j=1,2,\ldots,n_b$$

Fig. 3.2 Types of OBFs selected for model identification

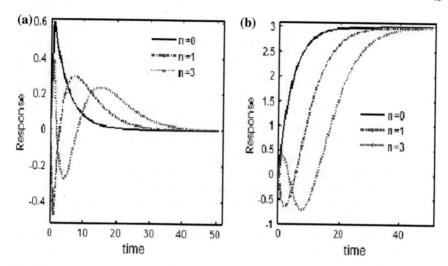

Fig. 3.3 Responses of the first three Laguerre Filters with pole: **a** Impulse Response, and **b** Step Response

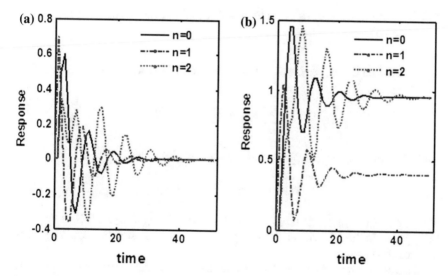

Fig. 3.4 Responses of the first three Kautz Filters with pole $\xi_{1,2} = 0.6 \pm j0.6$: **a** Impulse Response, and **b** Step Response

degrees $n_b > 0$. In 1997, Ninnes and Gustafson [9] came up with a unified approach by showing that the equation given in Fig. 3.2d makes up a complete orthogonal set with $\xi = \{\xi_j, j = 1, 2, \ldots\}$ signifying arbitrary poles inside the unit circle and in conjugate pairs. Latter in 2008, Toth [10] revealed that GOBF in fact resides to a bigger set of OBFs referred as Takanaki-Malmquist Functions (TMF). GOBFs have been chosen over the others for a system having scattered poles.

As partially stated, one limitation of LBF, KBF and GOBF is that the time delay is not considered. After the contribution by Finn et al. [11], however, the time delay is included by putting some of the poles at the origin. The resulting orthogonal set is named Markov-OBF by Patwardhan and Shah [12], Fig. 3.2e. Note that the three OBFs, including the pulse transfer functions, can be constructed from Markov-OBF. One drawback of using OBF is the pre-requirement on the dominant pole or poles of the system and time delay in case of using Markov-OBF. In Sect. 5.6 of the thesis, a method to iteratively estimate the dominant poles and time constants will be demonstrated.

3.2.2 Construction of Nonlinear Models in the Framework of OBFs

After including OBFs, the black-box model for an LTI system can be expressed as:

$$y(p) = \sum_{i=0}^{n} \theta_i g_i(q, \xi) + H(q, \theta)e(p) \tag{3.7}$$

while the nonlinear counter part takes the form as given by (3.8).

$$y(p) = f(g(q, \xi); \theta) + e(p) \tag{3.8}$$

where, $g(q, \xi) = [g_0(q, \xi)u(p) \ldots g_n(q, \xi)u(p)]^T$; $\xi = [\xi_1 \quad \xi_2 \ldots \xi_n]^T$ is vector of OBF poles. In the LTI models, model parameters can be estimated applying Least Squares (LS) method.

In terms of extent of studies, the most researched dynamic nonlinear model in the frame work of OBFs is the Volterra Systems [13–15]. In Volterra models, polynomial expansion of the input-output data is the foundation leading to the formulation of suitable models. A Volterra model has the advantage that it has a nonlinear structure, it is linear in parameters, and the model behaves stable for open-loop stable systems. However, Volterra models require model parameters in order to achieve the higher accuracy. In fact it is not common to see Volterra models with model orders higher than two because of the high computational burden. In the following, we would like to look at other alternative nonlinear models.

Even though Volterra models are well studied techniques, recently the use of Artificial Neural Networks (ANN) and Fuzzy Systems (FS) in the frame work of

OBF is introduced. In the research works by Parker and Tumma [16], Wray and Green [17], Marmarelis and Zhao [18], Liu et al. [19], Alataris et al. [20], we find detail studies about the conditions under which Volterra models could be developed from neural network approach. It was shown by Back and Tsoi [21] that ANN incorporating Laguerre filters can approximate a Volterra model to a certain degree of accuracy. In the same year, 1996, Sentoni et al. [22] also explored the possibility of developing a nonlinear model from the use of ANN with single hidden layer and Laguerre filters. After a test on control of a binary distillation column, they have concluded that the model using the stated approach is indeed effective. Besides, they mentioned that time delay in the system can be considered by making adjustments on the input to the Laguerre filters. Convergence and generalization characteristics of OBF-ANN based models are studied by Balestrino et al. [23] and Abrahantes Vazquez et al. [24], respectively. In the work of Diwanji et al. [25], we see the application of Laguerre functions and ANN based nonlinear models to the design of a model predictive controller for a single spool gas turbine. The model is a result of direct adaptation of Weiner model structure [26], Fig. 3.5. The whole idea was to perform feasibility study and at the end they came to notice that the proposed approach is indeed better than NARMAX model, which is traditionally known problematic in terms of fixing the model structure and model orders. It is worth noting that, this is the only work we came across in the area of power plants.

Now, we look into the models related to fuzzy systems. Sbarbaro and Johansen [27] are, to our knowledge, the first to demonstrate a nonlinear model that incorporates Laguerre filters and at the same time posses the characteristics of fuzzy modelling. They used operating region dependent local approximations linked by weighting parameters that are reflections of fuzzy sets. In the same year and inspired by Schram et al. work, Nelles [28] proposed a similar model but the model trained by LOLIMOT algorithm. Then latter in 1999, Oliveira et al. [29] reported a fuzzy relational model in the framework of OBFs. There are drawbacks regarding the models. First, they assumed equally spaced cluster centres. Second, spread terms are set equal to half of the distance between two adjacent centres. Third, they used a fixed

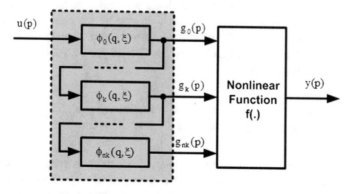

Fig. 3.5 Nonlinear model structure with the inclusion of OBFs [26]

pole of $\xi = 0.7$. None of these are realistic. In 2002, Campello and Amaral [30] published more general realization of the Oliveiras approach in which a state-space local model is assumed in the rules of the OBF-Fuzzy model. In their study, they used the technique to design a controller for a polymerization reactor. In the following year, they also extended the approach to hierarchical fuzzy models [31]. While testing the possibility of training the models by LS and GA algorithms, they also studied the performance of the resulting models as part of a Model Predictive Control (MPC) system for an ethanol production plant [32]. In fact we find also the use of GA for training OBF-Fuzzy model investigated by Medeiros et al. [33]. Unlike the other method, however, they used a fitness function based on Akaki Information Criterion (AIC). Knowing that GA approach may result in high computational time, Machado et al. [34] considered product space clustering to deal with the estimation of model parameters. The last two approaches are analysed using a data from magnetic levitation system. Recently, that is in 2009, Alci and Asyali [35] showed the training of a Laguerre function based fuzzy model by Levenberg Marquardt algorithm. For the analysis of the performance and generalization features of the approach, they applied the method to Box-Jankins gas furnace data and forced Van der Pol oscillator [36].

In the present section, the different model identification techniques with and without the inclusion of OBFs are discussed. In the next section, theoretical background for neural networks and fuzzy systems will be presented.

3.3 Neural Network and Fuzzy Systems

Actual systems are nonlinear resulting from variable operating conditions, coupled governing equations and complex energy exchange. In the regression approaches, models for a nonlinear system are developed assuming that the system works in the vicinity of design point and modelling the case by an approximate local linear models. Even then, any assumed regression model leads to the need to apply nonlinear optimization techniques for the measurement noise creates coupled set of equations.

In nonlinear model identification, the techniques from Computational Intelligence (CI) are often considered suitable. This is evidence by the brief review made in Chap. 2. The likes of neural networks, fuzzy systems, evolutionary computing and swarm intelligence make up this group. The next section provides details about neural networks and fuzzy systems.

3.3.1 Artificial Neural Networks

In a biological neural system, neurons or nerve cells are massively interconnected capable of exchanging signals to each other. A neuron consists of a cell body, dendrites, and an axon, Fig. 3.6. The connection in the neural system, referred as the synapse, is between the axon of one neuron to the dendrites of another neuron. The

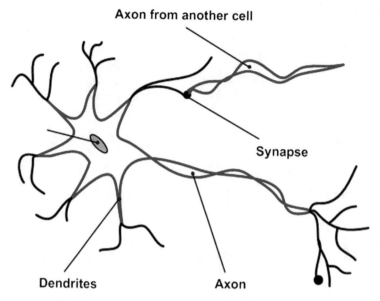

Fig. 3.6 Illustration of a biological neuron

signal in the nerves systems propagate from dendrites to the cell body, from the cell body to the axon. For the axon is connected to the dendrites, the same signal goes to all the linked dendrites. Interesting feature of the cell body is that, the signal heads to the other cell if and only if the cell body fires. The firing is commonly conceived as accompanied by altering the received signal or transmitting it as it is.

Artificial Neuron (AN) is a simplified model or representation of the biological neuron, Fig. 3.7. The inputs to the AN are multiplied by the corresponding weights and the sum is evaluated by a function referred to as activation function to excite the next connected neuron. In an artificial neural network, the ANs are connected in layers to define a specific type of ANN. The number of layers in a model could be just two or higher. For example, in Auto-Associative Neural Networks (AANN) [37] that performs self-mapping, there are five layers with the middle layer called bottle-neck layer [37]. The network was used in gas turbines [38–40] for fault detection and diagnosis.

The common activation functions are: linear function, sigmoid function, hyperbolic tangent function, radial basis function and wavelet function, Fig. 3.8. Based on the type of activation function, one finds such neural network as Multi-Layer Perceptron (MLP), Radial Basis Function (RBF) neural networks, Wavelet Neural Network (WNN), etc.

The performance of an ANN model is not only dependent on the number of layers and the kind of activation function. Equally important is the connection architecture. In general, the ANN could be featured by an architecture that is either fully connected or hierarchically connected. In the fully connected design, a neuron is connected

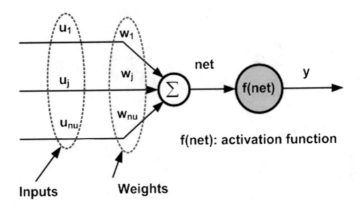

Fig. 3.7 Illustration of an artificial neuron

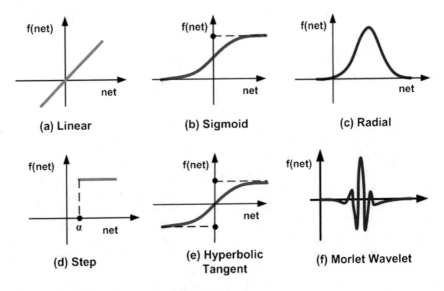

Fig. 3.8 Types of neural network activation functions

to other neurons and to itself. In the hierarchical design, a neuron in one layer is connected to neurons in the foregoing and subsequent layer only. If one assumes a neuron in the 1st layer, it gets input from the environment and the output goes to neurons of the following layer. As such, there is no connection between neurons in the same layer as well as no self connection, Fig. 3.9.

Neural networks are also classified in terms of signal directions as feed forward (Fig. 3.10a) and recurrent neural networks (Fig. 3.10b). In the feed forward design, a signal to a neuron comes from either the environment or from a neuron in the previous layer. And, an output from a neuron goes to either the environment or to the next neurons. In recurrent design, feedback signals from neurons in the very right

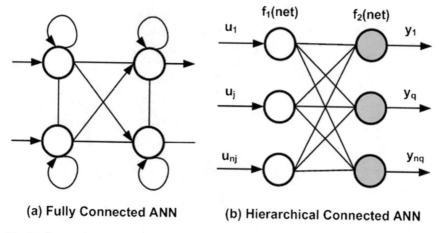

Fig. 3.9 Types of ANN connections: **a** Fully Connected ANN, and **b** Hierarchical Connected ANN

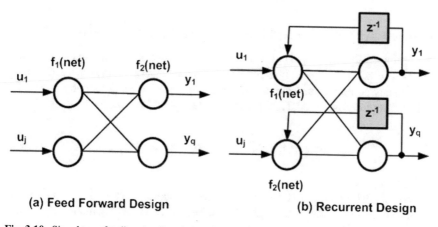

Fig. 3.10 Signal transfer direction based classification of ANN

or output layer are used as an input to neurons in the first layer. As compared to the feed forward design, recurrent ANN possesses elements of memory.

It is well proven that a multilayer feed forward ANN with a nonlinear activation function is capable of approximating any nonlinear function [41–43]. As stated at the outset, plants like cogeneration and district cooling are nonlinear in their operating characteristics. In fact, it is for this reason that we focused on nonlinear system identification techniques. ANNs can be trained easily for the activation functions are differentiable. There are plenty of learning algorithms that make ANN much attractive for model identification. In the following, the two commonly used ANN designs will be considered for further discussion.

Fig. 3.11 Feed forward
multilayer ANN or FFM-NN
structure

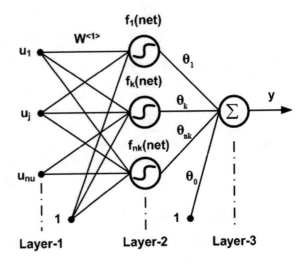

Layer-1   Layer-2   Layer-3

Feed Forward Multilayer Artificial Neural Network (FFM-ANN)

According to Cybennko [41], a two layer neural network with nonlinear activation
function (Fig. 3.11) sigmoid function or hyperbolic tangent has the power of approximating any nonlinear function. Considering input $u \in R^{nu}$ and output $y \in R^{ny}$, the
output from the ANN model and for $p - th$ data point is governed by (3.9).

$$y^p = 1.\theta_0 + \sum_{k-1}^{nk} f_k^p(\mathbf{u}^p; [\mathbf{b}^{<1>} \quad \mathbf{W}^{<1>}])\theta_k \tag{3.9}$$

where, $f_k^p(.)(k = 1, \ldots, nk)$: is the activation function; $\theta_k(k = 1, \ldots, nk)$: is the
model parameters or network weights for the second layer; $\mathbf{W}^{<1>}$: matrix of network
weights for the first layer.

Equation (3.9) can be extended to the whole data set of length N_d. In that case,
the network output can be expressed as

$$\mathbf{y} = \tilde{\mathbf{F}}_{\mathbf{MLP}}.\boldsymbol{\theta} \tag{3.10}$$

where, $\tilde{\mathbf{F}}_{\mathbf{MLP}} = [\mathbf{1}_{N_d \times 1} \quad \mathbf{F}_{\mathbf{MLP}}]$; $\mathbf{y} = [y(1)y(2) \ldots y(N_d)]^T$; $\boldsymbol{\theta} = [\theta_0 \theta_1 \ldots \theta_{nk}]^T$,
and $\mathbf{b}^{<1>} = [b_{1,1} b_{k,1} b_{nk,1}]$. The matrices are given by:

$$\mathbf{F}_{\mathbf{MPL}} = \begin{bmatrix} f_1(1) & f_k(1) & \cdots & f_{nk}(1) \\ f_1(2) & f_k(2) & \cdots & f_{nk}(2) \\ \vdots & \vdots & \ddots & \vdots \\ f_1(N_d) & f_k(N_d) & \cdots & f_{nk}(N_d) \end{bmatrix}, \text{ and }$$

$$\mathbf{W}^{<1>} = \begin{bmatrix} w_{1,1} & w_{k,1} & \cdots & w_{nk,1} \\ w_{1,2} & w_{k,2} & \cdots & w_{nk,2} \\ \vdots & \vdots & \ddots & \vdots \\ w_{1,nu} & w_{k,nu} & \cdots & w_{nk,nu} \end{bmatrix}$$

Simplicity of FFM-ANN structure and availability of different learning algorithms have made this network the most widely used in the area of model prediction and pattern classification. The use of FFM-ANN for FDD can be found in the area of gas turbines [44, 45], steam turbines and boilers [46], CHP [47, 48] and nuclear power plants [49–53].

Radial Basis Function Artificial Neural Network (RBF-ANN)

The architecture of RBF-ANN consists of three layers: input layer, hidden layer and output layer. The structure for a Multiple Input and Single Output (MISO) system is shown in Fig. 3.12. The activation function for node k in the hidden layer is a Gaussian function expressed as

$$f_k(\mathbf{u}^p; \mathbf{c}_k, \sigma_k) = exp\left[-\frac{1}{2}\{(\frac{u_1 - c_{k,1}}{\sigma_{k,1}})^2 + (\frac{u_2 - c_{k,2}}{\sigma_{k,2}})^2 + \cdots + (\frac{u_{nu} - c_{k,nu}}{\sigma_{k,nu}})^2\}\right]$$

$$(3.11)$$

where, $\sigma(k = 1, \ldots, nk)$: the standard deviation; $\mathbf{c}_k$: the centre vector. The output from the network for $p - th$ data is calculated as:

$$y^p = 1.\theta_0 + \sum_{k=1}^{nk} f_k((u)^p; \mathbf{c}_k, \sigma_k)\theta_k$$

Fig. 3.12 Radial basis function neural network or RBF-ANN architecture

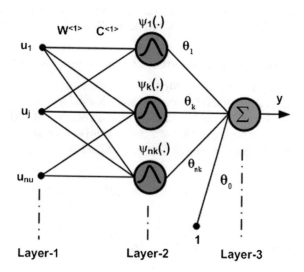

If stacked over the whole data set, then

$$\mathbf{y} = \tilde{\mathbf{F}}_{\mathbf{RBF}} \cdot \boldsymbol{\theta} \tag{3.12}$$

where, $\tilde{\mathbf{F}}_{\mathbf{RBF}} = [\mathbf{1}_{N_d \times 1} \quad \mathbf{F}_{\mathbf{RBF}}]$,

$$\mathbf{C}^{<1>} = \begin{bmatrix} c_{1,1} & c_{2,1} & \cdots & c_{nk,1} \\ c_{1,2} & c_{2,2} & \cdots & c_{nk,2} \\ \vdots & \vdots & \ddots & \vdots \\ c_{1,nu} & c_{2,nu} & \cdots & c_{nk,nu} \end{bmatrix}, \text{ and}$$

$$\Sigma = \begin{bmatrix} \sigma_{1,1} & \sigma_{2,1} & \cdots & \sigma_{nk,1} \\ \sigma_{1,2} & \sigma_{2,2} & \cdots & \sigma_{nk,2} \\ \vdots & \vdots & \ddots & \vdots \\ \sigma_{1,nu} & \sigma_{2,nu} & \cdots & \sigma_{nk,nu} \end{bmatrix}$$

The forms of $\mathbf{F}_{\mathbf{RBF}} \in R^{N_d \times (n_\theta+1)}$ and $\boldsymbol{\theta} \in R^{n_\theta}$ in (3.12) are similar to those defined for (3.10).

Training of radial basis function based ANN network is done in two steps:

- Clustering the input-output data space for calculating the cluster centres that define $\sigma_k \in R^{n_u}$ and $\mathbf{C}_k \in R^{n_u}$, and
- Training of θ_k values between hidden layer and output node using suitable algorithm. Least square is the easiest approach.

RBF-ANN presents such advantages as easily passing the local minima during training and the training is insensitive to the order of pattern presentation. RBF-ANN, however, are slow in training and they may need more hidden neurons for function approximation as compared to feed forward neural networks. Universal function approximation capabilities of RBF-ANN is well demonstrated in the reports of Park and Sandberg [54], and Chen [55]. Application of a probabilistic RBF-ANN to fault detection in cooling section of a coal power plant is investigated by Munoz and Sanz-Bobi [56]. A RBF-ANN trained by GA and used as OSD in nuclear power plants can be found in [57].

Normalized Radial Basis Function Artificial Neural Network

Figure 3.13 indicates NRBF-ANN structure [3]. The first two layers are similar to those for RBF-ANN. The additions are the normalization in layer-3. Basically, the normalization makes it more similar to a probabilistic RBF-ANN [56]. The activation function for node-k in the hidden layer is a Gaussian function. The output from the network and for $p - th$ data is described by (3.13).

$$y^p = \sum_{k=1}^{nk} \varphi_k \theta_k \tag{3.13}$$

where, $\varphi_k^p = \dfrac{f_k^p(\mathbf{u}^p; \mathbf{c}_k, \sigma_k)}{\sum\limits_{k=1}^{nk} f_k^p(\mathbf{u}^p; \mathbf{c}_k, \sigma_k)}$, with $\sum\limits_{k=1}^{nk} \varphi_k^p = 1$

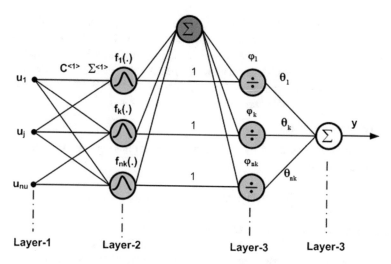

Fig. 3.13 Normalized radial basis function neural network (NRBF-ANN) structure

The form of $f_k^p(\mathbf{u}^p; \mathbf{c}_k, \boldsymbol{\sigma}_k)$ is as given by (3.11). If the model in (3.13) is stacked over the whole data, it forms into

$$\mathbf{y} = \phi.\boldsymbol{\theta} \tag{3.14}$$

where, $\phi = \begin{bmatrix} \varphi_1(1) & \varphi_k(1) & \cdots & \varphi_{nk}(1) \\ \varphi_1(2) & \varphi_k(2) & \cdots & \varphi_{nk}(2) \\ \vdots & \vdots & \ddots & \vdots \\ \varphi_1(N_d) & \varphi_k(N_d) & \cdots & \varphi_{nk}(N_d) \end{bmatrix}$

NRBF-ANN has good interpolation ability as compared to RBF-ANN [3]. So far the use of NRBF-ANN is limited to dynamic models only. Equations (3.10), (3.12) and (3.14) are similar in form. This condition will be exploited latter to formulate approximate expressions for confidence interval calculations.

3.3.2 Fuzzy Systems and Fuzzy Modelling

In the classical set theory belongingness of an element to a given set is described by a membership value of either 0 (if it doesnt belong to the set) or 1 (if it does). In real life, things are not this exact. For example, temperature high or low, hot or cold has vagueness in the description. Humans understand this easily while a machine finds it difficult to differentiate between hot and cold or high and low.

Fuzzy set and fuzzy logic allow the description of concepts or ideas with approximate reasoning. In fuzzy sets, an element can belong to two sets at the same time with partial membership in both of them. The membership value of an individual could take a value from [0, 1]. As such, fuzzy set allows reasoning with uncertain concepts to infer new concepts, with degree of certainty linked to each concept.

Application wise, fuzzy systems have been applied in control systems [58], model identification [3] and fault detection and diagnosis [59]. Fuzzy methods are tested as fault detection tools in the area of gas turbines [60, 61]. In the following sections, basic concepts of fuzzy sets and fuzzy logic are elaborated.

Fuzzy Inference System

The fuzzy method is defined by an inference system having sets of rules. General form of a fuzzy rule is

"if antecedent(s) then consequent(s)"

For a dynamic system, many of these rules are combined in a certain way defining knowledge base of the fuzzy reasoning system. While simulating, the fuzzy sets in the antecedent and consequents are combined using fuzzy logic operators (see Appendix A). In addition to the knowledge base, fuzzy rule-based reasoning systems are featured by fuzzification, inferencing and defuzzification blocks. The connection between the main blocks is shown in Fig. 3.14.

In the fuzzification block, the inputs are mapped to fuzzy membership values using appropriate fuzzy membership functions. Commonly used fuzzy functions are triangular, trapezoidal, logistic, and Gaussian, Fig. 3.15. In the inference block, the fuzzy values are combined on the bases of the rules predefined and saved in the knowledge base. The output from the inference system is the result from the defuzzification block in which fuzzy values from the rules involved in the process are aggregated following a preset method.

Fuzzy Modelling

In a nut shell, fuzzy modelling may be considered as the procedure of describing a system with sets of fuzzy rules. There are three classes of widely used fuzzy models:

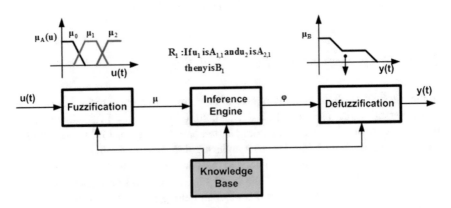

Fig. 3.14 Fuzzy rule based reasoning system

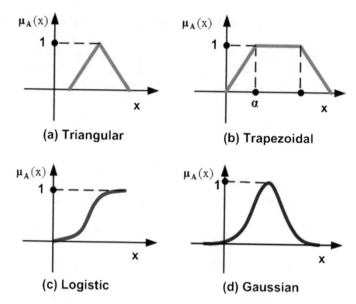

Fig. 3.15 Fuzzy rule based reasoning system

(i) Linguistic fuzzy models, (ii) fuzzy relational models, and (iii) Takagi-Sugeno-Kang (TSK) models [3]. Formulating a model using one of these methods involves estimating a suitable model structure and corresponding model parameters. For a Multiple Input and Multiple Output (MIMO) system, the linguistic fuzzy model is constructed by sets of rules that are expressed in the form of (3.15).

$$R_k : if\ u_1\ is\ A_{1k}\ and\ u_2\ is\ A_{2k}\ and\ \dots\ u_j\ is\ A_{j,k}\ \dots\ and\ ..\ u_{nu}$$
$$is\ A_{nu,k}\ then\ z_1\ is\ B_{1k}\ and\ y_2\ is\ B2k\ and\ \dots\ z_{nz}\ is\ B_{nz,k} \quad (3.15)$$

where, $u_j (j = 1, 2, \dots, nu)$: input linguistic variables; $z_q (q = 1, 2, \dots, nz)$ output linguistic variables; $A_{j,k}$ and $B_{q,k}$: fuzzy rules input and output membership functions, respectively. Construction of the linguistic fuzzy model requires the steps of deciding the number of rules and selection of suitable membership functions. The number of rules can be determined by clustering the input-output data. The choice on the type of membership function is dependent on how good the model fits the data and simplicity of the function. Fuzzy relational models for a system with two fuzzy sets A and B is specified by (3.16).

$$\mathbf{B} = \mathbf{A} \circ \mathbf{R} \quad (3.16)$$

where, $\circ$ is the fuzzy operator, Appendix A. The identification process is simply the estimation of $\mathbf{R}$ based on the input-output data simplified by fuzzy sets A_i and B_i,

$(i = 1, 2, \ldots, n)$. This approach in the framework of OBFs has been investigated by Oliveira et al. [29].

The third model, TSK model, for a Multiple Input Single Output (MISO) system is often described by rule of the type given by (3.17). As compared to the linguistic fuzzy model, this model has a general function in the consequent part.

$$R_k : if \ u_1 \ is \ A_{1k} \ and \ u_2 \ is \ A_{2k} \ and \ \ldots \ u_j \ is \ A_{j,k}$$
$$\ldots and \ .. \ u_{nu} \ is \ A_{nu,k} \ then \ z_k \ f_k(u_1, u_2, u_2, \ldots, u_n u) \tag{3.17}$$

where, $R_{nk}(k = 1, 2 \ldots, nk)$: rules in the model; $u_j(j = 1, 2, \ldots nu)$: crisp input variables; $A_{k,j}$: fuzzy subsets of the fuzzy variable u_j; z_k: output from k-th rule of the system. For given input variables, the model output is calculated as

$$y_k = \varphi_1 z_1 + \varphi_2 z_2 + \cdots + \varphi_k z_k + \cdots + \varphi_{nk} z_{nk} \tag{3.18}$$

where, φ_k is the rule validity index and is given by $\varphi_k = \frac{\mu_k}{\mu_1 + \mu_2 + \cdots + \mu_{nk}}$, for $k = 1, 2, \ldots, nk$

With μ_k signifying truth value of the antecedent part of the k-th rule calculated as

$$\mu_k = \prod_{j=1}^{nu} \mu_{A_{k,j}}, (u_j)$$

where, $\mu_{k,j}(u_j)$ is membership function of the fuzzy set $A_{k,j}$. The construction of TSK model entails determination of fuzzy membership parameters using appropriate clustering algorithms or optimization techniques. This step is the same for both linguistic model and TSK model. The membership functions after clustering define antecedent parts of the rule base. The parameters corresponding to the consequent part in the TSK model can be estimated from input-output data applying global optimization algorithms. In an alternative design, the fuzzy models can be structured into a neural network and trained as ANN model. In the following sections, discussions are made on neuro-fuzzy methods commonly applied for model identification.

3.4 Neuro-Fuzzy Model Identification

There are several ways to combine fuzzy systems with neural network [62]. One way is to adapt the fuzzy model in the neural network structure. The combined approach has the advantage of easy training of the model from input-output data. Experts knowledge can be used to define preliminary rules and membership functions for the model.

The neuro-fuzzy model contains the three steps in fuzzy reasoning process: fuzzification, inference and defuzziffication. In the inference section, the if-then part realizes the rule activation level. The remaining part of the rule realizes the calculation of the model output. In the following three different designs of neuro-fuzzy models are elaborated. All the designs possess identical antecedent part while their consequent parts are different.

3.4.1 Neuro-Fuzzy Singleton

NF model in the form of singleton outputs are expressed as [3]:

$$R_k : if \ u_1 \ is \ A_{1k} \ and \ u_2 \ is \ A_{2k} \ and \ \dots \ u_j \ is \ A_{j,k}$$
$$\dots \ and \ .. \ u_{nu} \ is \ A_{nu,k} \ then \ fault \ y = \alpha_k \tag{3.19}$$

where, $u_j (j = 1, 2 \dots, nu)$: crisp values of the inputs to the model; $A_{j,k}$: membership function of u_j with membership value of $\mu_{A_{j,k}}(u_j)$, $\alpha_k (k = 1, 2 \dots, nk)$: output (singleton). Figure 3.16 demonstrates NF singleton model structure for a Multiple Input and Single Output (MISO) system. It assumes a Gaussian membership function given by (3.20).

$$\mu_{A_{j,k}}(u_j) = exp\left(-\frac{1}{2}(\frac{u_j - c_{jk}}{\sigma_{k,j}^2})^2\right) \tag{3.20}$$

The model is realized by the processes in the four layers. In layer-1, the inputs are fuzzified to the corresponding fuzzy sets. The number of fuzzy sets for an input coincides with the number of rules in the model. The second layer realizes the fuzzy operator (see Appendix A) on the antecedent parts of the rules. The activation value of the rules in this layer are calculated as

$$\alpha_k = \prod_{j=1}^{nu} \mu_{j,k}(u_j) \tag{3.21}$$

In layer-3, validity function for each layer are calculated from the expression

$$\varphi_k = \frac{\alpha_k}{\alpha_1 + \alpha_2 + \dots + \alpha_{nk}} \tag{3.22}$$

And finally, the network output is calculated by the formula

$$y = \varphi_1 \theta_1 + \varphi_2 \theta_2 + \dots + \varphi_{nk} \theta_{nk} \tag{3.23}$$

where, $\theta_k (k = 1, 2 \dots, nk)$: weight in layer-4 that corresponds to the singletons. The model suggested in Fig. 3.16 is featured by n_k number of rules. In other models, the number of rules is pretty large for all the nodes corresponding to a specific input are combined to all other input nodes in defining the rule base. The NF singleton model is the base for formulating the higher order models.

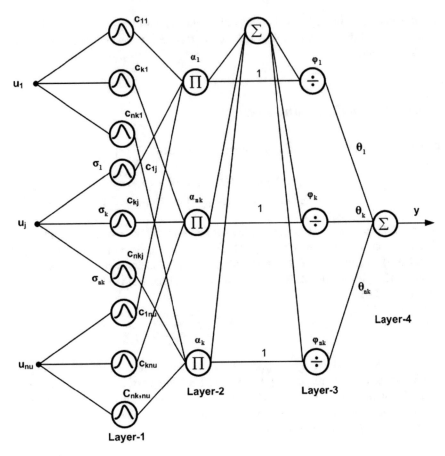

Fig. 3.16 NF singleton model for a MISO system

3.4.2 Takagi Sugeno Kang (TSK) Based Neuro-Fuzzy Network

Many authors use Neuro-Fuzzy (NF) approach to benefit from learning ability of neural networks and transparent or linguistic knowledge representation capability of fuzzy method. NF is good when limited information is available to model the system, first principle models are difficult to obtain and if the system is featured by nonlinear behavior. There are two frequently used types of NF approaches Mamdani type and Takagi Sugeno Kang (TSK) method. The TSK is suitable to describe a nonlinear system as a combination of piecewise linear models. Instead of using a singleton in the rule consequent, it employs linear in parameter models, Fig. 3.17.

Just like the fuzzy singleton model, the first step in TSK based NF model identification is clustering the input-output data and define Membership Functions (MF).

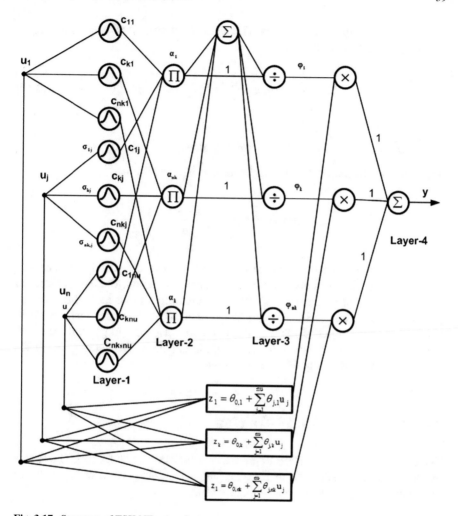

Fig. 3.17 Structure of TSK NF network structure

Latter, each input is fuzzyfied using the MF. Rules are then generated that signify local linear approximation of the nonlinear relation.

For a system having inputs $u_j (j = 1, 2 \ldots, nu)$ and output y, the kth fuzzy rule corresponding to the pth data resembles:

$$R_k^p : If\ u_1^p\ is\ \mu_{A,k}(u_j^p)\ and\ u_j^p\ is\ \mu_{A,k}(u_j^p)\ and \ldots and\ u_{nu}^p$$

$$is\ \mu_{A,k}(u_{nu}^p)\ then\ z_k^p = \theta_{k,0} + \sum_{j=1}^{nx} u_j^p \theta_{k,j} \tag{3.24}$$

If t-norm operator (see Appendix A) is applied for the antecedent part, the aggregate output for p^{th} data will be

$$y^p = \sum_{k=1}^{nk} \varphi_k^p (\theta_{k,0} + \sum_{j=1}^{nu} u_j^p \theta_{k,j}) = \sum_{k=1}^{nk} \varphi_k^p z_k^p \qquad (3.25)$$

where, $\varphi_k^p = \alpha_k^p / \sum_{k=1}^{nk} \alpha_k^p$ is the rule validity index; $\alpha_k^p = \prod_{j=1}^{nu} \mu_{A,k}(u_j^p)$ is the rule aggregate membership value. The k-th membership function is

$$\mu_{A_{j,k}}(u_j) = exp\left(-\frac{1}{2}\left(\frac{u_j - c_{jk}}{\sigma_{k,j}^2}\right)^2\right)$$

where, σ_k^2 and $c_{k,j}$ are the variance and center values, respectively of cluster k. Assuming that cluster centers and variances are pre determined, (3.25) can be stacked for $N - d$ number of data and written in short form so that local linear model parameters could be identified using global or local least squares technique. NF approach has been applied for fault detection and diagnosis in gas turbines [63], nuclear reactors [64], flow control valves [65], and steam generators [66]. In the following, the learning algorithms are presented.

3.5   Neuro-Fuzzy Model Training

There are several algorithms for training a Neuro-Fuzzy model [3, 59]. The choice on the particular type depends on the input-output data and purpose of the model. In the following only three of them are highlighted.

3.5.1   Least Squares Based Learning Algorithms

In the LS based estimation algorithm, the parameters in the antecedent part of the NF model are first calculated either by distance based clustering algorithms or heuristic input-space decomposition technique [3]. The later is advantages as it involves no optimization procedures. After the antecedent parameters are determined, LS technique is usually used to determine consequent parameters. If all the parameters are determined at ones from overall regression matrix and corresponding output data, the least square technique is referred as Global Least Squares (GLS) [3]. In the following the Local Linear Model Tree Algorithm (LOLIMOT) as developed by Nelles and Isermann [67] will be considered.

3.5.2 Estimation of Rules Consequent Parameter

In local linear approximations, the consequent parameters are determined by considering the consequent parameter optimization problem as independent problems. As such, parameters of a rule are estimated separately neglecting the interaction between local models. Latter, the output of the NF model for a given input is determined mostly by the local linear model while the others are almost inactive.

The total number of consequent parameters in the NF model is $n_v = n_k(n_u + 1)$. In the GLS technique, all these parameters are optimized at once. In LOLIMOT approach, n_k independent local linear optimizations are performed to estimate $(n_u + 1)$ parameters of each NF rule. The parameter vector for each of the rules $k = 1, 2, \ldots, n_k$ is

$$\boldsymbol{\theta}_k = \begin{bmatrix} \theta_{0,k} & \theta_{1,k} & \cdots & \theta_{nu,k} \end{bmatrix}^T$$

And, the corresponding regression matrix is given by

$$\mathbf{R}_k = \begin{bmatrix} 1 & u_1^{(1)} & u_2^{(1)} & \cdots & u_{n_u}^{(1)} \\ 1 & u_1^{(2)} & u_2^{(2)} & \cdots & u_{n_u}^{(2)} \\ \vdots & \vdots & & \ddots & \vdots \\ 1 & u_1^{(N_d)} & u_2^{(N_d)} & \cdots & u_{nu}^{(N_d)} \end{bmatrix} \tag{3.26}$$

where, N_d stands for the number of data samples. Because the elements of the $\mathbf{R}_k$ do not rely on k, the regression matrix of all local linear models $k = 1, 2 \ldots, n_k$ are identical. Applying Weighted Least Squares (WLS), for a given input-output data, the locally linear model parameters are calculated as

$$\theta = (\mathbf{R}_k{}^T \mathbf{Q}_k \mathbf{R}_k)^{-1} \mathbf{R}_k{}^T \mathbf{Q}_k \mathbf{t} \tag{3.27}$$

where, $\mathbf{t} = [t^{(1)} \quad t^{(2)} \quad \cdots \quad t^{(N_d)}]^T$ is the actual output; $\mathbf{Q}_k = \varphi_k^T \varphi_k$ is a weighting diagonal matrix with $\varphi_k = diag([\varphi_k^{(1)} \varphi_k^{(2)} \cdots \varphi_k^{(p)} \cdots \varphi_k^{(N_d)}])$. The local model output $\hat{\mathbf{y}}_k = \mathbf{R}_x \boldsymbol{\theta}_k$ is solely valid in the region where the corresponding validity function φ_k is close to one. This regions is the region close to the centre of φ_k. In LOLIMOT algorithm, a parameter called model validity index I_k is defined based on φ_k and the model error $e^{(p)} = (t^{(p)} - \hat{y}^{(p)})$ for the purpose of measuring the quality of a local linear model. Important step of the algorithm is to partition the input space relying on the validity index and retrain the model until the preset error limit or maximum number of partitions is reached. A local model with higher value of validity index is considered for partition. The equation for the validity index is

$$I_k = \sum_{p}^{N_d} \varphi_k^{(p)} (t^{(p)} - \hat{y}^{(p)})^2 \tag{3.28}$$

In the sequel, the input space decomposition technique that defines the LOLIMOT is presented.

Input Space Decomposition Algorithm

The LOLIMOT algorithm as developed by Nelles and Isermann [67] involves two major calculation loops. In the outer loop, a check is performed to make sure that the set prediction error limit is reached. In the inner loop, partitions along each dimension are carried out followed by Local Linear Model (LLM) identification. The validity index as defined by (3.28) is applied to decide where to make partitions for next estimation. In summarized form, LOLIMOT algorithm entails the following steps.

(a) Start the estimation with a simple regression model. That is, with

$$y^{(p)} = 1.\theta_0 + \sum_{j=1}^{n_u} u_j^{(p)} \theta_j$$

And calculate the model parameters by LS method. The regression matrix is given by (3.26) that refers to just one rule.

(b) Divide the input space along each input dimension and into two halves. For each partitioned region calculate the hyper rectangle centre and the corresponding spread terms. According to Nelles [3], the latter term along a certain input dimension can be assumed to have one-third of the hyper rectangle size along the same dimension. The centre and spread terms will be used to formulate fuzzy membership functions for each hyper rectangle. The partitioning process for a model having two inputs and a single output is demonstrated in Fig. 3.18.

(c) Create a LLM for each newly formed hyper rectangle. At this stage (3.27) will be used to estimate the model parameters.

(d) Calculate the global model prediction error for each model and select the one with minimum prediction error.

(e) Within the model selected in step-(d), calculate a local validity index I_k for each of the local models using (3.28) and find the worst LLM that corresponds to a maximum value of the validity index.

(f) Check if the minimum prediction error calculated so far is smaller than what is specified as optimization termination parameter. If it is satisfied, then stop the optimization.

(g) If the prediction error limit is not reached, then partition the hyper rectangle with the maximum validity index and repeat steps (b) to (f).

The LOLIMOT algorithm has been applied in [66] for training NF models intended for fault detection and diagnosis in nuclear reactors. Nelles [28], who took the credit for the algorithm, investigated the use of the algorithm for training NF-OBF models. The algorithm is fast as compared to derivative based and random search global algorithms. The drawback in the algorithm is the assumption on the calculation of the spread terms from the hyper rectangles. As such, it presupposes a heuristic term that is equal to 1/3.

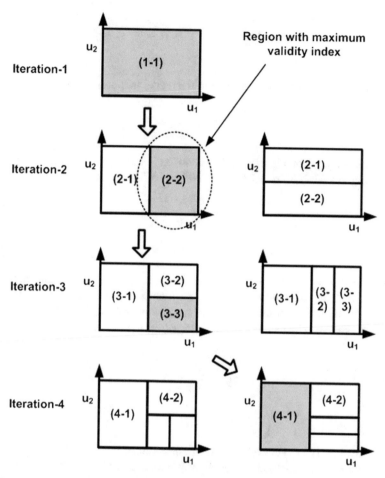

Fig. 3.18 Steps of input space decomposition leading to LOLIMOT algorithm [67]

3.5.3 Gradient Descent Learning Algorithm

Parameters of the nonlinear NF model, (3.25) can be determined by error back-propagation algorithm, conjugate gradient method, evolutionary computing or swarm intelligence techniques. In fact, the same set of algorithms is applicable to train models represented by (3.10), (3.12), (3.14), and (3.23). The easiest of all is back-propagation. However, it needs derivatives of the error function (3.29) with respect to the model parameters. The initialization can be done by data clustering followed by least squares algorithm. After initialization, equations of the following form are used to iteratively determine the model parameters.

Assuming that sum of squares of the error between the model $\hat{y}$ and the actual data t is expressed as

$$\mathbf{V}(\hat{\boldsymbol{\theta}}) = \sum_{p=1}^{Nd} (t^p - \hat{y}^p)^2 = (\mathbf{t} - \hat{\mathbf{y}})^T (\mathbf{t} - \hat{\mathbf{y}}) \qquad (3.29)$$

The derivative of $\mathbf{V}(\hat{\boldsymbol{\theta}})$ with respect to the model output is

$$\frac{\partial \mathbf{V}(\hat{\boldsymbol{\theta}})}{\partial \hat{\mathbf{y}}} = -2(\mathbf{t} - \hat{\mathbf{y}}) \qquad (3.30)$$

In back-propagation algorithm, the following iterative equations can be adopted for training the FFM-ANN, RBF-ANN, NF, and NF-OBF models.

$$\Delta \boldsymbol{\theta}_{i+1} = -\eta \frac{\partial \mathbf{V}}{\partial \boldsymbol{\theta}} + \alpha \Delta \boldsymbol{\theta}_i \qquad (3.31)$$

$$\boldsymbol{\theta}_{i+1} = \boldsymbol{\theta}_i + \Delta \boldsymbol{\theta}_{i+1} \qquad (3.32)$$

where, $\boldsymbol{\theta} = [w_1 w_2 \dots w_n \quad \rho_1 \rho_2 \dots \rho_n \dots \quad c_1 c_2 \dots c_n]^T$, $\boldsymbol{\theta}_i$: the parameters at step 1, $\Delta \boldsymbol{\theta}_i$: parameter adaptation at step i, α: the momentum coefficient, η: the learning rate, $\frac{\partial \mathbf{V}}{\partial \boldsymbol{\theta}}$: the derivative of the sum of squares of the model error with respect to the model parameters. The value of the derivative over the data set is

$$\frac{\partial \mathbf{V}}{\partial \boldsymbol{\theta}} = \sum_{p=1}^{Nd} \left(\frac{\partial \mathbf{V}^p}{\partial \boldsymbol{\theta}} \right) \qquad (3.33)$$

where, $\frac{\partial \mathbf{V}^p}{\partial \boldsymbol{\theta}} = \left[\left(\frac{\partial V^p}{\partial \theta_1} \right) \dots \left(\frac{\partial V^p}{\partial \theta_k} \right) \left(\frac{\partial V^p}{\partial \theta_{nk}} \right) \right]$

If the spread terms and cluster centers are determined by clustering algorithms or heuristic input space decomposition algorithm, the procedure reduces only to calculating the parameters in the consequent parts of the rules. The terms required in (3.33) and corresponding to neural network and Neuro-Fuzzy models are elaborated in Appendix B.

Derivative based learning algorithms have been used in the research work by Ruz-Hernandez [68], Wenli et al. [69], Rakhshani et al. [70], and Hwang [49]. In [35], a particular type of derivative based training algorithm Levenberg Marquardt is used to design a Laguerre filter based fuzzy model. One drawback of the methods is the possibility of converging to the local minimum. In the following, a global optimization algorithm is considered.

3.5.4 Particle Swarm Optimization

Particle Swarm Optimization (PSO) is a nature inspired global optimization algorithm that is free from any derivative of the objective function, (3.33). It was first proposed by Kennedy and Eberhart [71]. They created the method to simulate behaviors of birds flocking together in search for food. In PSO algorithm, some numbers of candidate solutions are first created, each named as particle. The set formed by the particles is regarded as a swarm. The length of the particle corresponds to the number of model parameters to be estimated N_θ. Assuming that the number of particles is designated by N_p, the swarm can be expressed in matrix form as

$$\mathbf{S} = [\mathbf{S}_1 \quad \mathbf{S}_2 \quad \ldots \quad \mathbf{S}_{Np}] \tag{3.34}$$

where, $\mathbf{S}_k = [\mathbf{S}_{1,k} \quad \mathbf{S}_{2,k} \quad \ldots \quad \mathbf{S}_{N_p,k}]$ for $k = 1, 2, \ldots, N_p$ is a vector that relates to the particles position in the solution space $R^{N_d \times N_p}$. In fact the position and the name particle are interchangeable. To start the optimization, a velocity vector is also defined for each particle. We represent the corresponding velocity matrix as

$$\mathbf{V} = [\mathbf{V}_1 \quad \mathbf{V}_2 \quad \ldots \quad \mathbf{V}_{Np}] \tag{3.35}$$

where, $\mathbf{V}_k = [v_{1,k} \quad v_{2,k} \quad \ldots \quad v_{N_\theta,k}]^T$ for $k = 1, 2, \ldots, N_p$ signifies velocity of each particle in the solution space. At the start of the optimization, each particle is initialized with a random position $\mathbf{S}_k(i = 0)$ and velocity $\mathbf{V}_k(i = 0)$. In PSO algorithm, the particle in the solution space moves to a new position at a velocity that is dynamically adjusted as a function of current position of the particle, own previous best position and best position discovered by the whole swarm in the latest generation. For each new position, the objective function $\mathbf{f}(\mathbf{S}_k)$ is evaluated to identify the best over the whole swarm and in terms of the chosen criteria. For each generation, fundamental steps of the PSO algorithm – (3.36) and (3.37) are used to update the velocity $\mathbf{V}_k(t)$ and position $\mathbf{S}_k(t)$ vectors. In (3.36), the second term stands for learning from its own translation experience. The third term, however, refers to the correction to the velocity update resulting from information sharing with other neighbors. Particles local and global best positions, respectively, are represented by $\mathbf{S}_k^{ib}(t)$ and $\mathbf{S}_k^{ib}(t)$.

$$\mathbf{V}_k(t + 1) = \alpha \mathbf{V}_k(t) + \eta_1.diag(\mathbf{r}_1(t)).(\mathbf{S}_k^{ib}(t) - \mathbf{S}_k(t)) + \eta_2.diag(\mathbf{r}_2(t)).(\mathbf{S}^{gb}(t) - \mathbf{S}_k(t)) \tag{3.36}$$

$$\mathbf{S}_k(t + 1) = \mathbf{S}_k(t) + \chi \mathbf{V}_k(t + 1) \tag{3.37}$$

The term α is called the inertia weight. Its meaning is equivalent to the momentum term in the back propagation algorithm. $\mathbf{r}_1(t)$ and $\mathbf{r}_2(t)$ vectors define stochastic part of the PSO algorithm. These vectors are random values uniformly distributed in the range of 0 to 1. η_1 and η_2 are constants that decide how far a particle can go in a certain direction; they are also called acceleration coefficients. χ is termed the constriction factor. For convergence reasons, the velocities of the particles are limited to

$$v_{k,j} = \begin{cases} v_{k,j} & if \ v_{k,j} \leq v_{min} \\ v_{max} & if \ v_{k,j} \geq v_{max}, \ j = 1, 2, \ldots, N_p \ and \ k = 1, 2, \ldots N_d \\ v_{k,j} & otherwise \end{cases} \quad (3.38)$$

with v_{max} assumed to be half of the search range. The inertia weight α may be kept constant throughout the total generation. However, linear (3.39) or exponential (3.40) variations could be assumed that facilitates the local convergence.

$$\alpha = \alpha_{min} + (1 - i/i_{max})(\alpha_{max} - \alpha_{min}) \quad (3.39)$$

$$\alpha = \alpha_{max} - exp(-\beta/t) \quad (3.40)$$

where, α_{min} and α_{max} are minimum and maximum value, respectively, of the inertia term; i_{max} is the maximum number of generations until the optimization terminates; β is a constant to control the rate of decrease of the inertia term or learning rate. Assuming $N_\theta = 2$, a vector diagram related to (3.36) and (3.37) are indicated in Fig. 3.19.

The original PSO algorithm did not include the inertia term and constriction factor. They were added by Eberhart and Shi [72], and Clerc and Kennedy [73], respectively, after their discovery that the two terms may help improve convergence. Following [73], the constriction factor χ is a function of η_1 and η_2 as given by (3.41).

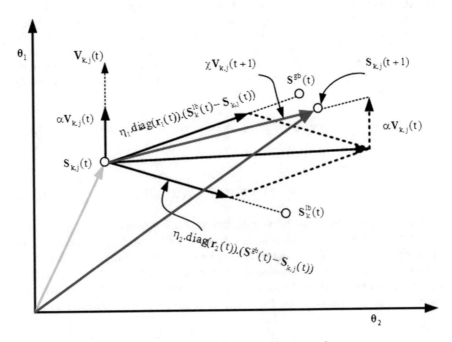

Fig. 3.19 Vector diagram representation of the PSO algorithm, $N_\theta = 2$

$$\chi = \frac{2}{\left|2 - \varphi - \sqrt{\varphi^2 - 4\varphi}\right|}, \; \varphi = \eta_1 + \eta_2, \; \varphi > 4 \qquad (3.41)$$

The value of χ is often considered equal to 0.729 that corresponds to φ set to 4.1. There are also other proposed approaches focused at improving the PSO algorithm. In the report by Yang et al. [74], the velocity terms are modified replacing η_1 and η_2 by $(1 + \varepsilon_1)/3$ and $(1 + \varepsilon_2)/3$, respectively. Where ε_1 and ε_2 are small positive numbers. Huang et al. [75], however, added a fourth term $a \times (\mathbf{r}_3(t) - 0.5)$ to the equation for the new velocity instead of changing the values of the acceleration terms. In the new term, a is a small number while $\mathbf{r}_3(t)$ possesses the property of $\mathbf{r}_1(t)$ and $\mathbf{r}_2(t)$. Different from the first two modification, Liu et al. [76] on the other hand adopted acceleration calculation to trace the rate of decrease of velocity of a particle and perturb it if it tends to go to zero while the solution is still not converging. PSO converges fast to the global optimum but has difficulties in fine tuning solutions, Porto et al. [77]. Besides, setting a proper inertia weight and suitable velocities are common problems that could cause the algorithm unable to converge.

Group Based Particle Swarm Optimization (GB-PSO)

In the PSO method discussed so far, a particle is free to move in the solution space and its locations are updated based on globally best particle position and the best previous local position of the particle itself, (3.36). In the following, we propose a slightly different arrangement with the objective of enhancing convergence speed.

Suppose we have groups of birds flocking together in search for food with one group communicating with the other and each bird in a group getting information from its own group members and other adjacent groups. If we further assume that a particle in a group is assigned best position performance measure quantified as Root Mean Squared Error (RMSE) of the error function that is a function of its location, (3.41)–(3.43) can be applied for updating next local best position of the particle.

For best performing particle in a group, the performance measure is featured by

$$\mathbf{RMSE}_k^*(t) = min \begin{pmatrix} \mathbf{RMSE}_{k,1}(t), \mathbf{RMSE}_{k,1}(t), \ldots, \\ \mathbf{RMSE}_{k,j}(t), \ldots, \mathbf{RMSE}_{k,nj}(t) \end{pmatrix} \qquad (3.42)$$

In line with (3.41), the current local best for the group is given by:

$$\mathbf{S}_k^{lb}(t) = \mathbf{S}_{k,j}(t), \; iff \; \mathbf{RMSE}_{k,j} = \mathbf{RMSE}_k^*(t) \qquad (3.43)$$

Then, for the next iteration, the local best is updated by:

$$\mathbf{S}_k^{lb}(t+1) = \begin{cases} \mathbf{S}_k^{lb}(t), iff \; \mathbf{RMSE}_k(t) < \mathbf{RMSE}_k(t-1) \\ \mathbf{S}_k^{lb}(t-1), iff \; \mathbf{RMSE}_k(t) \geq \mathbf{RMSE}_k(t-1) \end{cases} \qquad (3.44)$$

The rest of the calculation follows (3.36) and (3.37). As compared to the conventional PSO algorithm, the proposed modification allows fast convergence with more flexibility to search for global solutions.

3.6   A New Optimization Algorithm Based on Mating Behavior of Simien Jackal

Yet another nature inspired algorithm could be designed making use of breeding characteristics of Simien Jackal or Ethiopian Wolf [78]. Simien Jackals are endemic animals in Ethiopia. While other jackals are widely distributed, Ethiopian Wolves only live in the highlands where rodents are abundant. They are known living in packs of 2–13. Due to the relative ease to mate with domestic dogs, they are also thought as more closely related to grey wolves and coyotes than African canids.

In a Jackals pack, only the dominant female can breed. The other females either disperse or become floaters. Ascendance to a breeding status by a new female occurs only after the death of the dominant female. During mating season, the dominant female was observed accepting only the dominant male in the females pack. Besides, the female was seen having 70% of the total copulation practiced with males outside the pack regardless of male status. In some instances, a dominant female jackal was also seen mating with domestic dogs. The remaining 30.

Now, to make use of the breeding behavior as optimization method, we would like to make the following assumptions:

- We assume that a pack is having all females and there is one dominant female that can control the characteristics of the rest of the pack. We represent the dominant female by $\mathbf{S}_k^{lb}$ and the rest of the jackals in the pack by $\mathbf{S}_{k,j}$.
- In passing genes from one generation to another, we assume that the new generation in a pack gets genes from the dominant female and a jackal outside the pack. We distribute the normalized copulation number as $\eta \in [0, 1]$ and $(1 - \eta)$ to the dominant male in the pack and the male outside tha pack, respectively. To account for the random mating with a domestic dog or a stranger jackal, we intend to assign a mutation probability of $\gamma_m \in [0, 1]$.
- We further assume that in all the copulation that take place between the dominant female in a pack and a male outside the pack, $\alpha \in [0, 1]$ portion of $(1 - \eta)$ goes to the female gene that is found globally best while the remaining goes to the rest of the genes anticipated for copulation.

Now, based on the stated assumptions, the updating equation for a jth jackal in kth pack is:

$$\mathbf{S}_{k,j}(t + 1) = (1 - \alpha)(1 - \eta) \times \mathbf{S}^{gb}(t) + \alpha(1 - \eta) \times [\mathbf{S}_k^{lb}(t)]^* + \eta \times \hat{\mathbf{S}} \quad (3.45)$$

$$\hat{\mathbf{S}} = \begin{cases} \mathbf{S}_k^{lb}(t), & if \ \gamma > \gamma_m \\ \tilde{\mathbf{S}}, & if \ \gamma \le \gamma_m \end{cases} \tag{3.46}$$

where, $[\mathbf{S}_k^{lb}]^*$ is the jackal selected randomly but known to the current generation; $\tilde{\mathbf{S}}$ is the jackal randomly created based on probability of mutation γ_m. $\mathbf{S}_k^{lb}(t)$ of a given pack is calculated according to Eqs. (3.41)–(3.43). In the proposed algorithm, from now on called as Jackal (JCK) algorithm, there is no velocity calculation. Typical value for η is 0.3, which matches with the 30

$$\alpha_k^{lin} = \left(\frac{n_k - k^* - 1}{n_k - 1} \right) \tag{3.47}$$

$$\alpha_k^{exp} = exp(-(\alpha_k^{lin} - 1)) \tag{3.48}$$

where, $k^* \in [1, 2, \ldots, n_k]$ and is selected randomly.

3.7   Summary

In this chapter, the basic concepts behind nonlinear model identification were addressed. Neuro-Fuzzy method designed in the framework of OBFs was selected to model subsystems of the CDCP under dynamic conditions. The methods are basic inputs to the design of the fault detection system that will be presented in Chap. 5. Hereunder, the core points from the current chapter are summarized.

- Developing a first principle model for CDCP systems is difficult attributed to the need to know all the design point data that is hardly available. In case of having high fidelity models, the simulation is often not fast due to high order of complexity in the model. Hence, for practical design of FDD systems, we ought to resort to black-box models. This approach allows not only capturing input-output relations based on historical data but also simulated data.
- In the CDCP, the subsystems can be modelled either under dynamic conditions or steady state conditions. The second case is common when our aim is to monitor performance deterioration, which in most cases is a gradual process. Hence, in a situation like that, it may not be necessary to rely on time based description. For monitoring safety critical parameters, however, dynamic models have to be considered. To this end, it is a must to construct the models based on past inputs and past outputs.
- For the CDCP systems, two kinds of models steady state and dynamic have to be developed. In both cases, the nonlinear characteristics that are partly caused by variable operating conditions need be included. Regarding this, the current chapter addressed artificial neural networks, fuzzy systems and their combination.
- FIR based dynamic system identification demonstrated higher model orders for a fairly accurate prediction. Recently, the use of OBFs was found effective in

reducing the model orders significantly. The current chapter discussed possible model structures in the framework of OBFs and their characteristics.

- System poles and time delay are major inputs to OBF based designs. Once the parameters are known, a specific type of OBF can be adopted that better approximates dynamics of the system. In the current chapter, an approach was suggested to for the design of OBF based models.
- Three different algorithms Least Squares (LS), Back-Propagation (BP) and Particle Swarm Optimization (PSO) are explained. The choice of the three is based on their frequent use and considering the fact that the LS could not be applied alone if GOBF is considered in the model and the poles are not known ahead. Apparently, the back-propagation algorithm can be trapped at the local minimum. The best option to search for global minimum estimation error is then to use metaheurstics algorithms where PSO is one of them. As compared to evolutionary programming, genetic algorithm and evolutionary strategies, PSO does not need any chromosome formulation, parent selection, mutation and cross-over operations. The current chapter elaborated the assumptions and main relations behind each method.

Finally, while presenting the basic concepts required for the development of an improved FDD system, it showed to what extent OBF has been used in nonlinear system identification. The GB-PSO and JCK algorithms are two additional contributions of the current chapter. The flowcharts behind the two algorithms are provided in Figs. D.9 and D.10, respectively. By and large, it can be said that the presentation has met the objective set at the start of the chapter, which mainly concentrates on nonlinear model identification. The outcome of this chapter is crucial to the design if the fault detection and diagnosis system, which is the main objective of in the thesis.

References

1. L, Ljung. (1999). *System identification: Theory for the user.*, Prentice Hall Information Series.
2. Sderstrm, T., & Stoica, P. (1988). *System identification.*, Prentice Hall International Series in Systems and Control Engineering Englewood Cliffs: Prentice-Hall, Inc.
3. Nelles, O. (2001). *Nonlinear system identification: From classical approaches to neural networks and fuzzy models.* Berlin: Springer.
4. Wahlberg, B. (1991). System identification using Laguerre models. *IEEE Transactions on Automatic Control, 36*, 551–562.
5. Brinker, A. C. D. (1995). Meixner-like functions having a rational z-transform. *International Journal of Circuit Theory and Applications, 23*, 237–246.
6. Belt, H. J. W. (1997). *Orthonormal bases for adaptive filtering.* Eindhoven: Technische Universiteit.
7. Wahlberg, B. (1994). System identification using Kautz models. *IEEE Transactions on Automatic Control, 39*, 1276–1282.
8. Heuberger, P. S. C., Van den Hof, P. M. J., & Bosgra, O. H. (1995). A generalized orthonormal basis for linear dynamical systems. *IEEE Transactions on Automatic Control, 40*, 451–465.
9. Ninness, B., & Gustafsson, F. (1997). A unifying construction of orthonormal bases for system identification. *IEEE Transactions on Automatic Control, 42*, 515–521.
10. Toth, R. (2008, December). Modeling and identification of linear parameter-varying systems, an orthonormal basis function approach. Ph.D. dissertation, Delft University of Technology.

11. Finn, C. K., Wahlberg, B., & Ydstie, B. E. (1993). Constrained predictive control using orthogonal expansion. *Journal of AIChE, 39*, 1810–1826.
12. Patwardhan, S. C., & Shah, S. L. (2005). From data to diagnosis and control using generalized orthonormal basis filters. Part I: Development of state observers. *Journal of Process Control, 15*, 819–835.
13. Boyd, S., & Chua, L. (1985). Fading memory and the problem of approximating nonlinear operators with Volterra series. *IEEE Transactions on Circuits and Systems, 32*, 1150–1161.
14. Ogunfunmi, T. (2007, September, 5). *Adaptive nonlinear system identification: The Volterra and Wiener model approaches.* New York: Springer Science and Business Media.
15. Seretis, C., & Zafiriou, E. (1997). Nonlinear dynamical system identification using reduced Volterra models with generalised orthonormal basis functions. In *Proceedings of the American Control Conference* (Vol. 5, pp. 3042–3046).
16. Parker, R. E., & Tummala, M. (1992). Identification of Volterra systems with a polynomial neural network. In *IEEE International Conference on Acoustics, Speech, and Signal Processing, ICASSP-92* (Vol. 4, pp. 561–564).
17. Wray, J., & Green, G. (1994). Calculation of the Volterra kernels of non-linear dynamic systems using an artificial neural network. *Biological Cybernetics, 71*, 187–195.
18. Marmarelis, V. Z., & Zhao, X. (1997). Volterra models and three-layer perceptrons. *IEEE Transactions on Neural Networks, 8*, 1421–1433.
19. Liu, G. P., Kadirkamanathan, V., & Billings, S. A. (1998). On-line identification of nonlinear systems using Volterra polynomial basis function neural networks. *Neural Networks, 11*, 1645–1657.
20. Alataris, K., Berger, T., & Marmarelis, V. (2000). A novel network for nonlinear modeling of neural systems with arbitrary point-process inputs. *Neural Networks, 13*, 255–266.
21. Back, A. D., & Tsoi, A. C. (1996). Nonlinear system identification using discrete laguerre functions. *Journal of Systems Enginnering, 6*, 194–270.
22. Sentoni, G., Agamennoni, O., Desages, A., & Romagnoli, J. (1996). Approximate models for nonlinear process control. *Journal of AIChE, 42*, 2240–2250.
23. Balestrino, A., & Caiti, A. (2000). Approximation of Hammerstein/Wiener dynamic models. In *Proceedings of the IEEE-INNS-ENNS International Joint Conference on Neural Networks, IJCNN 2000* (Vol. 1, pp. 70–74).
24. Abrahantes Vazquez, M., & Agamennoni, O. E. (2001). Approximate models for nonlinear dynamical systems and their generalization properties. *Mathematical and Computer Modelling, 33*, 965–986.
25. Diwanji, V., Godbole, A., & Waghode, N. (2006). Nonlinear model predictive control for thrust tracking of a gas turbine. In *IEEE International Conference on Industrial Technology, ICIT 2006* (pp. 3044–3048).
26. Wiener, N. (1949). *Extrapolation, interpolation, and smoothing of stationary time series.* New York: MIT and Wiley.
27. Sbarbaro, D., & Johansen, T. A. (1997). Multiple local Laguerre models for modelling nonlinear dynamic systems of the Wiener class. In *IEE Proceedings Control Theory and Applications* (Vol. 144, 375–380).
28. Nelles, O. (1997). Orthonormal basis functions for nonlinear system identification with local linear model trees-LOLIMOT. In *IFAC Proceedings Volumes* (Vol. 30(11), pp. 639–44).
29. Oliveira, G. H. C., Campello, R. J. G. B., & Amaral, W. C. (1999). Fuzzy models within orthonormal basis function framework. In *IEEE International Fuzzy Systems Conference Proceedings, FUZZ-IEEE '99* (Vol. 2, pp. 957–962).
30. Campello, R. J. G. B., & Amaral, W. C. (2002). Takagi-Sugeno fuzzy models within orthonormal basis function framework and their application to process control. In *Proceedings of the 2002 IEEE International Conference on Fuzzy Systems, FUZZ-IEEE'02* (pp. 1399–1404).
31. Campello, R. J. G. B., Von Zuben, F. J., Amaral, W. C., Meleiro, L. A. C., & Maciel Filho, R. (2003). Hierarchical fuzzy models within the framework of orthonormal basis functions and their application to bioprocess control. *Chemical Engineering Science, 58*, 4259–4270.

32. Campello, R. J. G. B., Meleiro, L. A. C., & Amaral, W. C. (2004). Control of a bioprocess using orthonormal basis function fuzzy models. In *Proceedings of the 2004 IEEE International Conference on Fuzzy Systems* (Vol. 2, pp. 801–806).

33. Medeiros, A. V., Amaral, W. C., & Campello, R. J. G. (2006). GA optimization of generalized OBF TS fuzzy models with global and local estimation approaches. In *2006 IEEE International Conference on Fuzzy Systems* (pp. 1835–1842).

34. Machado, J. B., Amaral, W. C., & Campello, R. J. G. (2007). Design of OBF-TS fuzzy models based on multiple clustering validity criteria. In *19th IEEE International Conference on Tools with Artificial Intelligence, ICTAI 2007* (pp. 336–339).

35. Alci, M., & Asyali, M. H. (2009). Nonlinear system identification via Laguerre network based fuzzy systems. *Fuzzy Sets and Systems, 160*, 3518–3529.

36. Box, G. E. P., & Jenkins, G. M. (1976). *Time series forcasting and control* (Revised ed.). San Francisco: Holden-Day Inc.

37. Kramer, M. A. (1991). Nonlinear principal componenet analysis using autoassociative neural networks. *AIChE Journal, 37*, 233–243.

38. Ogaji, S., Sampath, S., Singh, R., & Probert, D. (2002). Novel approach for improving power-plant availability using advanced engine diagnostics. *Applied Energy, 72*, 389–407.

39. Lu, P. J., Zhang, M. C., Hsu, T. C., & Zhang, J. (2001). An evaluation of engine faults diagnostics using artificial neural networks. *Journal of Engineering for Gas Turbines and Power, 123*, 340–346.

40. Bourassa, M. A. J. (1999). *Autoassociative neural networks with an application to fault diagnosis of a gas turbine engine* (p. 248). Canada: Royal Military College of Canada (Canada).

41. Cybennko, G. (1989). Approximation by superpositions of a sigmoidal function. *Mathematics of Control, Signals, and Systems, 2*, 303–314.

42. Funahashi, K. I. (1989). On the approximate realization of continuous mappings by neural networks. *Neural Networks, 2*, 183–192.

43. Hornik, K., Stinchcombe, M., & White, H. (1989). Multilayer feedforward networks are universal approximators. *Neural Networks, 2*, 359–366.

44. Simani, S., Fantuzzi, C., & Spina, R. P. (1998). Application of a neural network in gas turbine control sensor fault detection. In *Proceedings of the 1998 IEEE International Conference on Control Applications* (Vol. 1, pp. 182–186).

45. Ogaji, S. O. T., & Singh, R. (2002). Advanced engine diagnostics using artificial neural networks. In *2002 IEEE International Conference on Artificial Intelligence Systems (ICAIS 2002)* (pp. 236–241).

46. Sreedhar, R., Fernandez, B., & Masada, G. Y. (1992). A neural network based adaptive fault detection scheme. In *Proceedings of the American Control Conference* (Vol. 5, pp. 3259–3263).

47. Arranz, A., Cruz, A., Sanz-Bobi, M. A., Ruz, P., & Coutio, J. (2008). DADICC: Intelligent system for anomaly detection in a combined cycle gas turbine plant. *Expert Systems with Applications, 34*, 2267–2277.

48. Fast, M., & Palm, T. (2010). Application of artificial neural networks to the condition monitoring and diagnosis of a combined heat and power plant. *Energy, 35*, 1114–1120.

49. Hwang, B. C. (1993). *Fault detection and diagnosis of a nuclear power plant using artificial neural networks*. Thesis: Simon Fraser University, Canada.

50. Keyvan, S. (2001). Traditional signal pattern recognition versus artificial neural networks for nuclear plant diagnostics. *Progress in Nuclear Energy, 39*, 1–29.

51. Embrechts, M. J., & Benedek, S. (1997). Identification of nuclear power plant transients with neural networks. In: *IEEE International Conference on Systems, Man, and Cybernetics, Computational Cybernetics and Simulation* (Vol. 1, pp. 912–916).

52. Weerasinghe, M., Gomm, J. B., & Williams, D. (1998). Neural networks for fault diagnosis of a nuclear fuel processing plant at different operating points. *Control Engineering Practice, 6*, 281–289.

53. Fantoni, P. F., & Mazzola, A. (1996). A pattern recognition-artificial neural networks based model for signal validation in nuclear power plants. *Annals of Nuclear Energy, 23*, 1069–1076.

54. Park, J., & Sandberg, I. W. (1991). Universal Approximation Using Radial Basis Function Networks. *Neural Computation, 3,* 246–257.
55. Chen, T., Chen, A. T., & Chen, R. (1995). Approximation capability to functions of several variables, nonlinear functionals and operators by radial basis function neural networks. *IEEE Transactions on Neural Networks, 6,* 904–910.
56. Muoz, A., & Sanz-Bobi, M. A. (1998). An incipient fault detection system based on the probabilistic radial basis function network: Application to the diagnosis of the condenser of a coal power plant. *Neurocomputing, 23,* 177–194.
57. Chun-ling, X., Jen-Yuan, C., Xiao-cheng, S., & Jing-min, D. (2008). Fault diagnosis of nuclear power plant based on genetic-RBF neural network. In *15th International Conference on Mechatronics and Machine Vision in Practice, M2VIP* (pp. 334–339).
58. Verbruggen, H. B., Zimmermann, H. J., & Babuska, R. (1999). *Fuzzy algorithms for control.,* International Series in Intelligent Technologies Boston: Kluwer Academic Publishers.
59. Korbicz, J., Koscielny, J. M., & Kowalczuk, Z. (2004). *Fault diagnosis: Models, artificial intelligence, applications.* Berlin: Springer.
60. Diao, Y., & Passino, K. M. (2004). Fault diagnosis for a turbine engine. *Control Engineering Practice, 12,* 1151–1165.
61. Ogaji, S. O. T., Marinai, L., Sampath, S., Singh, R., & Prober, S. D. (2005). Gas-turbine fault diagnostics: A fuzzy-logic approach. *Applied Energy, 82,* 81–89.
62. Ayoubi, M., & Isermann, R. (1997). Neuro-fuzzy systems for diagnosis. *Fuzzy Sets and Systems, 89,* 289–307.
63. Palade, V., Patton, R. J., Uppal, F. J., Quevedo, J., & Daley, S. (2002). Fault diagnosis of an industrial gas turbine using neuro-fuzzy methods. In *Proceedings of the 15th IFAC World Congress Barcelona* (pp. 2477–2482).
64. Zio, E., & Gola, G. (2006). Neuro-fuzzy pattern classification for fault diagnosis in nuclear components. *Annals of Nuclear Energy, 33,* 415–426.
65. Korbicz, J., & Kowal, M. (2007). Neuro-fuzzy networks and their application to fault detection of dynamical systems. *Engineering Applications of Artificial Intelligence, 20,* 609–617.
66. Razavi-Far, R., Davilu, H., Palade, V., & Lucas, C. (2009). Model-based fault detection and isolation of a steam generator using neuro-fuzzy networks. *Neurocomputing, 72,* 2939–2951.
67. Nelles, O., & Isermann, R. (1996). Basis function networks for interpolation of local linear models. In *Proceedings of the 35th IEEE Decision and Control* (Vol. 1, pp. 470–475).
68. Ruz-Hernandez, J. A., Sanchez, E. N., & Suarez, D. A. (2005). Neural networks-based scheme for fault diagnosis in fossil electric power plants. In *Proceedings of the IEEE International Joint Conference on Neural Networks IJCNN '05* (Vol. 3, pp. 1740–1745).
69. Wenli, Y., Lee, K. Y., Junker, S. T., & Ghezel-Ayagh, H. (2008). Fault diagnosis and accommodation system with a hybrid model for fuel cell power plant. In *Power and Energy Society General Meeting - Conversion and Delivery of Electrical Energy in the 21st Century* (pp. 1–8): IEEE.
70. Rakhshani, E., Sariri, I., & Rouzbehi, K. (2009). Application of data mining on fault detection ad prediction in Boiler of power plant using artificial neural network. In *International Conference on Power Engineering, Energy and Electrical Drives, POWERENG '09* (pp. 473–478).
71. Kennedy, J., & Eberhart, R. (1995). Particle swarm optimization. In *Proceedings of the IEEE International Conference on Neural Networks* (Vol. 4, pp. 1942–1948).
72. Eberhart, R. C., & Shi, Y. (2000). Comparing inertia weights and constriction factors in particle swarm optimization. In *Proceedings of the 2000 Congress on Evolutionary Computation* (Vol. 1, pp. 84–88).
73. Clerc, M., & Kennedy, J. (2002). The particle swarm - explosion, stability, and convergence in a multidimensional complex space. *IEEE Transactions on Evolutionary Computation, 6,* 58–73.
74. Yang, E., Erdogan, A. T., Arslan, T., & Barton, N. (2007). An improved particle swarm optimization algorithm for power-efficient wireless sensor networks. In *ECSIS Symposium on Bio-inspired, Learning, and Intelligent Systems for Security (BLISS 2007)* (pp. 76–82).

75. Huang, D. S., Li, K., Irwin, G., He, Q., & Han, C. (2006). An improved particle swarm optimization algorithm with disturbance term. In *Computational Intelligence and Bioinformatics* (Vol. 4115, pp. 100–108). Berlin: Springer.
76. Liu, D., Fei, S., Hou, Z., Zhang, H., Sun, C., Hu, W., et al. (2007). A new BP network based on improved PSO algorithm and its application on fault diagnosis of gas turbine. In *Advances in Neural Networks* (Vol. 4493, pp. 277–283). Heidelberg: Springer.
77. Porto, V., Saravanan, N., Waagen, D., Eiben, A., & Angeline, P. (1998). Evolutionary optimization versus particle swarm optimization: Philosophy and performance differences. In *Evolutionary Programming VII* (pp. 601–610). Heidelberg: Berlin.
78. Macdonald, D. W., & Sillero-Zubiri, C. (2004). *The biology and conservation of wild canids*. Oxford: Oxford University Press. (3 June 2010).

Chapter 4
Model Uncertainity, Fault Detection and Diagnostics

4.1 Introduction

The previous chapter has explained the concepts behind NF based model identification and how it relates to other models and the design in the framework of OBFs. It was stated that a good nonlinear model can be developed from plant operation data or a simulated output without knowing the model structure. However, the model alone is not enough for condition monitoring. In fact, the accuracy of a model is dependent on the estimated model parameters. In this regard, we may have one optimum parameter set out of many parameter sets all capable to characterize the system. In fault detector design, the knowledge of the whole set is critical as the fault detection and diagnosis system relies on model thresholds. In Sect. 4.2 of the chapter, the methods in the calculation of model uncertainity for linear in parameter models and nonlinear in parameter models, respectively, are explained. In the linear case, the equations for upper and lower prediction bounds are defined relying on iid and bounded error assumptions. In Sect. 4.3 the fault detection will be discussed while Sect. 4.4 is dedicated to the design of a fault diagnosis system that operates on bianry or fuzzy signals. Section 4.5 outlines summary of the chapter.

4.2 Model Uncertainty Calculation for Robust Fault Detector Design

The easiest way for calculating the confidence limit for a general parametric model is to use three-sigma limit. There are also other methods. Weyer et al. [1] used Cumulative Sum (CUSUM) detector for fault detection in heat exchangers. Hill and Minsker [2] applied Prediction Intervals (PI) calculated from recent historical data to detect streaming data anomalies. Similar application of the method can also be found in the work of Mencar [3]. Witczak et al. [7], and Marcin and Korbicz [4] adopted bounded error approach for GMDH based neural network to detect faults

© Springer International Publishing AG 2018
T.A. Lemma, *A Hybrid Approach for Power Plant Fault Diagnostics*,
Studies in Computational Intelligence 743,
https://doi.org/10.1007/978-3-319-71871-2_4

in valve actuators. In this thesis, linear and nonlinear models are treated separately, and model uncertainty equations are formulated relying on iid and bounded error assumption, respectively.

4.2.1 Linear in Parameter Models

The main feature of RBF-ANN, NRBF-ANN, Fuzzy Singleton and Fuzzy TSK model is that the technique of Least Mean Square (LMS) [5] can be applied for training the models if one assumes that certainty of the models are unaffected by nonlinear activation functions and membership function parameters. Even in FFM-ANN, the same idea works if one assumes that models uncertainty is less affected by all model parameters except those in the last layer. Membership Functions (MFs) are often determined by clustering methods, axes orthogonal partitioning or other optimization algorithms. Assuming the highlighted conditions hold, estimation of model parameters by LMS requires that the system output corresponding to pth input data is described by (4.1).

$$y^p = \Psi_p^T \theta^* + \varepsilon_p \tag{4.1}$$

where, $y^p \in R$ is the model output, $\theta^* \in R^{n_\theta}$ is vector of true parameters in the consequent part of the fuzzy rules, $\Psi_p \in R^{n_\theta}$ is the regression vector, and ε_p is the measurement and model error. Defining an approximation to the actual output as

$$\hat{y}^p = \Psi_p^T \hat{\theta} \tag{4.2}$$

where, $\hat{\theta} \in R^{n_\theta}$ is estimation related to the actual parameter, $\theta^* \in R^{n_d}$ estimate of the combined error term will be

$$\varepsilon_p = y^p - \hat{y}^p \tag{4.3}$$

The value of $\theta^* \in R^{n_d}$ can be determined applying batch methods like Maximum Likelihood (ML), Least Square Error (LSE) [5], or Minimum Mean Square Error (MMSE). In a different choice, one may also rely on a recursive algorithm that could be of Recursive Least Squares (RLS), Weighted Recursive Least Square (WRLS), Least Mean Square (LMS), Instrument Variable (IV) [6], or Kalman-Bucy Filter (KBF) [5].

Most of the mentioned algorithms are featured by the assumption that expectation and covariance of the error vector $\varepsilon \in R^{N_d}$ are zero and $\sigma^2 \mathbf{I}$, respectively. The first assumption emphasizes that the system is unaffected by deterministic or structural errors. And, the second assumption ensures that the model errors are considered to be random and represented by a stochastic model; as such the measurement noises are not correlated. Parameter estimations under these assumptions are common for it is fairly simple to execute. Nevertheless, this way of characterizing the errors is far from what practically happens. Even manufacturers of measuring devices provide working limits rather than statistical parameters. Hence, the better approach is to consider them as bounded but unknown. In fact, parameter estimation method under bounded error assumption and for a linear system is a well developed concept [8, 9]. In bounded error assumption, point wise bounds are considered known. If it is assumed that ε_p is the error for pth data with the corresponding bounds, then

$$(\varepsilon_p)_{min} \leq (\varepsilon_p) \leq (\varepsilon_p)_{max} \tag{4.4}$$

where, $(\varepsilon_p)_{min}$ and $(\varepsilon_p)_{max}$ are point-wise minimum and maximum bound, respectively, of the error. Now, (4.1)–(4.4) can be combined to formulate (4.6), which is an inequality equation. As such, for each data point, there are two inequalities the solution of which is any parameter vector $\hat{\theta} \in R^{n_\theta}$ that satisfies the inequalities. For a system having N_d number of data, the assumption results in $2 \times N_d$ inequalities.

$$y^p - (\varepsilon_p)_{min} \leq \hat{y}^p \leq y^p - (\varepsilon_p)_{max} \tag{4.5}$$

At this point, it can be said that bounded error assumption leads to a Linear Programming (LP) problem. If S is assumed to represent a strip in the parameter space $\Omega \in R^{n_\theta}$, bounded by two parallel hyper planes (4.6), the feasible parameter set corresponding to pth data set is

$$S^p = \{\theta \in R^{n_\theta} | y^p - (\varepsilon_p)_{min} \leq \hat{y}^p \leq y^p - (\varepsilon_p)_{max}\} \tag{4.6}$$

The intersection between all the strips forms a polyhedron, which is the solution space. It is well known that LP problems are computationally expensive, especially if the number of parameters to be estimated is large. For an LP problem, the optimum solution is decided based on a geometrical criterion. If the interest is on the estimation of uncertainty of the model as well, then the procedure will get even more involved. For reasons of simplifying the procedure, the actual shape of the solution space a polyhedron is often approximated by axes orthogonal shapes and special geometries like ellipsoid. Accordingly, there are LP problem solving algorithms called orthotopic algorithm, parallelotopic algorithm, interval methods [10], and ellipsoid algorithms [11]. Improved calculation method using Hopfield networks was proposed by Arruda et al. [12].

In the ellipsoid algorithm, which is a recursive method and often classified as set membership approach, the polyhedron is approximated by an ellipsoid. In the algorithm, successive ellipsoids are constructed containing all the feasible values of θ. For the pth measurement, the ellipsoid or feasible parameter set $\Omega(\hat{\theta}_p, \mathbf{M}_p)$ is given by (4.8).

$$\Omega(\hat{\theta}_p, \mathbf{M}_p) = \{\theta \in R^{n_\theta} | (\theta - \hat{\theta}_p)^T (\mathbf{M}_p)^{-1} (\theta - \hat{\theta}_p) \leq \sigma_p^2\} \tag{4.7}$$

Where, σ_p^2 is a nonzero scalar, $\hat{\theta}_p \in R^{n_\theta}$ is the center of the ellipsoid consistent with the first pth data sets and $\mathbf{M}^p$ is a positive definite matrix that represents the size and orientation of the ellipsoid. Intersection between the pth strip (4.7) and the ellipsoid is used to formulate the next feasible set and the procedure is repeated until the final data point is reached. The centre of the last ellipsoid signifies the resulting parameter set while the ellipsoid carries all possible solutions. Equation 4.9 expresses the relation between two successive ellipsoids. Details of the ellipsoid algorithms are presented in Sect. 4.2. As will be shown, an ellipsoid algorithm allows fairly simple calculation of model uncertainty intervals.

$$\Omega(\hat{\theta}_{p+1}, \mathbf{M}_{p+1}) = \Omega(\hat{\theta}_p, \mathbf{M}_p) \cap S^p \tag{4.8}$$

Model Uncertainty Estimation under iid Error Assumption

Under iid error assumption, the calculation of model threshold or confidence interval requires prior estimation of the model parameters and corresponding confidence regions. For LMS approach, the optimum parameter estimate is given by:

$$\hat{\theta} = [\boldsymbol{\Psi}^T \boldsymbol{\Psi}] \boldsymbol{\Psi}^T \mathbf{y} \tag{4.9}$$

Substitution of (4.1) into (4.10) and rearranging results in

$$\hat{\theta} = \theta^* + [\boldsymbol{\Psi}^T \boldsymbol{\Psi}] \boldsymbol{\Psi}^T \varepsilon \tag{4.10}$$

From (4.11), the expected value of θ is

$$\mathbf{E}\{\theta\} = \theta^* + [\boldsymbol{\Psi}^T \boldsymbol{\Psi}] \boldsymbol{\Psi}^T . \mathbf{E}\{\varepsilon\} \tag{4.11}$$

Because $\mathbf{E}\{\varepsilon\}$ is zero, $\mathbf{E}\{\theta\}$ will be just $\mathbf{E}\{\theta\} = \theta^*$. This shows that θ is unbiased. With the result that expectation of θ is θ^*, the parameters variance can then be stated as

$$Cov(\theta) = E\{\delta\delta^T\} \tag{4.12}$$

Where, $\delta = \hat{\theta} - \theta^*$. Inserting (4.11) into (4.12),

$$E\{\delta\delta^T\} = E\{[\boldsymbol{\Psi}^T\boldsymbol{\Psi}]^{-1}\varepsilon\varepsilon^T\boldsymbol{\Psi}[\boldsymbol{\Psi}^T\boldsymbol{\Psi}]^{-1}\}$$

Denoting $[\boldsymbol{\Psi}^T\boldsymbol{\Psi}]^{-1}$ by $\mathbf{M}$, the expectation equation reduces to

$$E\{\delta\delta^T\} = \mathbf{M}\boldsymbol{\Psi}^T E\{\varepsilon\varepsilon^T\}\boldsymbol{\Psi}\mathbf{M} \tag{4.13}$$

Using the assumption that measurement noise at different points is uncorrelated,

$$E\{\varepsilon\varepsilon^T\} = \sigma^2 \mathbf{I} \tag{4.14}$$

Applying (4.14) in (4.13), covariance of the parameter vector becomes

$$Cov(\hat{\theta}) = \sigma^2 \mathbf{M}(\mathbf{\Psi}^T \mathbf{\Psi})\mathbf{M} = \sigma^2 \mathbf{M} \tag{4.15}$$

From which, the confidence region for θ with a confidence level of $1 - \alpha$ is calculated as

$$\|\hat{\theta}_j - \theta_j\| = t_{\alpha, N_d - nm - 1}\sqrt{\hat{\sigma}^2 . M_{jj}} \, j = 1, \ldots, n_\theta \tag{4.16}$$

where, $t_{\alpha, N_d - n_\theta - 1}$ is percentage value of t-student distribution for $N_d - n_\theta - 1$ degrees of freedom. $t_{\alpha, N_d - n_\theta - 1}$ leaves $\alpha/2$ in the upper tail and $(1 - \alpha/2)$ in the lower tail of the distribution. M_{jj} signifies j-th diagonal element of the matrix $\mathbf{M} = [\mathbf{\Psi}^T \mathbf{\Psi}]^{-1}$. $\hat{\sigma}^2$ represents the unbiased estimate of the variance corresponding to the model error ε. $\hat{\sigma}^2$ is not known and hence has to be calculated from the available data.

$$\hat{\sigma}^2 = \frac{1}{N_d - n_\theta - 1} \sum_{p=1}^{N_d} \left(y^p - \hat{y}^p\right)^2 \tag{4.17}$$

Note that $\hat{\sigma}^2$ is assumed constant in the input space. The confidence region for θ with the confidence level of $(1 - \alpha)$ can also be written as

$$(\hat{\theta} - \theta)^{\mathbf{T}} \mathbf{\Psi}^{\mathbf{T}} \mathbf{\Psi}(\hat{\theta} - \theta)^{\mathbf{T}} \le (nm + 1) F_{\alpha, N_d - nm - 1}^{nm + 1} \tag{4.18}$$

where, $F_{\alpha, N_d - n_v - 1}^{n_v + 1}$ is the percentage value of F-distribution for $N_d - n_\theta - 1$ and $n_v + 1$ degrees of freedom. Calculation of variance can be extended to the output model. To this end,

$$E\{(\mathbf{y} - \hat{\mathbf{y}})(\mathbf{y} - \hat{\mathbf{y}})^T\} = E\{\mathbf{\Psi} \delta \delta^{\mathbf{T}} \mathbf{\Psi}^{\mathbf{T}} - 2\partial \delta^{\mathbf{T}} \mathbf{\Psi}^{\mathbf{T}} + \varepsilon \varepsilon^{\mathbf{T}}\} \tag{4.19}$$

Substituting $E\{\varepsilon\varepsilon^{\mathbf{T}}\}$ from (4.14) into (4.19), and assuming that is uncorrelated with δ,

$$Cov(\hat{\mathbf{y}}) = \sigma^2 (\mathbf{I} + \mathbf{\Psi} \mathbf{M} \mathbf{\Psi}^{\mathbf{T}}) \tag{4.20}$$

Finally, the prediction interval at a point $\mathbf{u}$ in the input space with a confidence level of $(1 - \alpha)$ is expressed as:

$$\|\hat{\mathbf{y}} - \mathbf{y}\|_\alpha = \mathbf{CI}_\alpha = t_{\alpha, N_d - n_v - 1}\{\sigma^2(\mathbf{I} + \mathbf{\Psi} \mathbf{M} \mathbf{\Psi}^{T})\}^{\frac{1}{2}} \tag{4.21}$$

Application of (4.21) can be found in [4].

Bounded Error Approach and Outer Bounding Ellipsoid Algorithm

The assumption that expectation of measurement errors is zero may not be realistic specially if there is no information on the characteristics of the errors. Besides, the assumption that structural errors are random while having a nonlinear model with limited measured data may not be acceptable. In scenarios like these, the suggested option is to resort to characterizing the errors as bounded between minimum and maximum limits. In fact, this suggestion is supported by the possibility of defining the bounds from knowledge about the system or using the technical information provided by the manufacturer about the measuring device.

In the bounded error model, a linear in parameter problem turns to a linear programming problem. Solution of the model equations will then lead to model parameter set estimation whose elements are compatible with the measurements, model structure and preset error bound. Under the assumption that (4.1) and (4.3) are satisfied, the point-wise error bounds $(\varepsilon)_{min}$ and $(\varepsilon)_{max}$ are given, $\|(\varepsilon)_{min}\|_2 = \|(\varepsilon)_{max}\|_2 = \gamma_p$ and, (4.15) describes the feasible parameter set S^p in the parameter space that is bounded by hyper planes $H_{p,1}$ and $H_{p,2}$.

$$S^P = \{\boldsymbol{\theta} \in R^{n_\theta} : |y_p - \boldsymbol{\psi_p^T}\boldsymbol{\theta}| \leq \gamma_p\}, \, p = 1, 2, \ldots, N_d \tag{4.22}$$

The equations for the hyper planes are

$$H_{p,1} = \{\boldsymbol{\theta} \in R^{n_\theta} | \boldsymbol{\psi_p^T}\boldsymbol{\theta} = y_p + \varepsilon_p\},$$

and

$$H_{p,1} = \{\boldsymbol{\theta} \in R^{n_\theta} | \boldsymbol{\psi_p^T}\boldsymbol{\theta} = y_p - \varepsilon_p\}$$

Each of the hyper planes partitions the parametric space into two half spaces. The half spaces form the feasible parameter set and this set is a monotone non-increasing sequence of sets having a polytopic shape [8]. Figure 4.1a indicates the feasible set for a model having only two model parameters and three data points. As shown in the figure, the resulting feasible set is a polyhedron. For the general case, the feasible set may be stated as intersections between all the half spaces, (4.23). Any parameter vector that falls inside the feasible region is considered as a valid parameter vector. Besides, the centre of the feasible region, in some geometrical sense, is often considered as the estimate of the true parameter vector $\boldsymbol{\theta}^*$.

$$S^P = \bigcap_{p=1}^{N_d} H_p^+ = \bigcap_{p=1}^{N_d} \left(H_{p,1}^+ \cap H_{p,2}^+\right) \tag{4.23}$$

where, $H_{p,1}^+ = \{\boldsymbol{\theta} \in R^{n_\theta} | \boldsymbol{\psi_p^T}\boldsymbol{\theta} \leq y_p + \varepsilon_p\}$, and $H_{p,1}^+ = \{\boldsymbol{\theta} \in R^{n_\theta} | \boldsymbol{\psi_p^T}\boldsymbol{\theta} \geq y_p - \varepsilon_p\}$. For the linear in parameter model, the exact shape of the parameter set is a polytope. For large data set, the polytope is quite complex. The way to minimize the computational time is to approximate the polytope by other geometries like ellipsoids [13], orthotops or parallelotopes.

Designating the enclosing or outer bounding approximation of the feasible set by Ω_p, recursive construction of the approximation may be written as an intersection between previous feasible set and the strip consistent with the current data set,

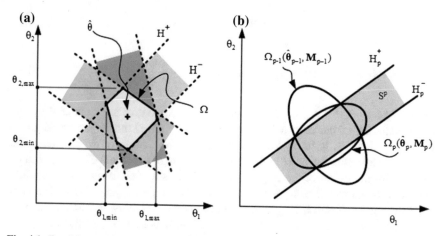

Fig. 4.1 Feasible parameter set ($n_\theta = 2$ and $N_d = 3$) and OBE Algorithm: **a** feasible parameter set, and **b** OBE algorithm

(4.24). In the following, the Optimum Bounding Ellipsoid (OBE) algorithm will be presented, Fig. 4.1b.

$$\Omega_p = \Omega_{p,1} \cap H_{p,1}^+ = \Omega_{p,1}^+ \cap H_{p,1}^+ \cap H_{p,2}^+ \tag{4.24}$$

In set membership theory, the approximation of the solution space for a parameter estimation problem by a circumscribing ellipsoid was first proposed by Schweppe [14]. After fourteen years since Schweppes work, Fogel and Huang [13] came up with the first Optimal Bounding Ellipsoid (OBE) algorithm featured by selective updating technique. The algorithm eliminates those data points having little or no effect on reducing the size of the ellipsoid. In 1987, Dasgupta and Huang [15] developed an improved version of the OBE algorithm. However, it was noted that the method minimizes a quantity that is not directly linked to the size of the ellipsoid and it lacks interpretability regarding the optimization criterion. After Dasguptas result, many authors published different versions of OBE. In the journal paper by Deller et al. [11], it was shown that all the algorithms that work by minimizing the size of the ellipsoid are developed on the bases of prior specification of point-wise error bounds and employing (4.25)–(4.29).

$$\varepsilon_p = y_p - \psi_\mathbf{p}^\mathbf{T}\theta_{p-1} \tag{4.25}$$

$$G_p = \psi_\mathbf{p}^\mathbf{T}\mathbf{M}_{p-1}\psi_p \tag{4.26}$$

$$\mathbf{M}_p = \frac{1}{\alpha_p}\left(1 - \frac{\beta_p\mathbf{M}_{p-1}\psi_p\psi_p^T}{\alpha_p + \beta_p G_p}\right)\mathbf{M}_{p-1} \tag{4.27}$$

$$\boldsymbol{\theta}_p = \boldsymbol{\theta}_{p-1} + \beta_p \mathbf{M}_p \psi_p \varepsilon_p \tag{4.28}$$

$$\sigma_p^2 = \sigma_p \sigma_{p-1}^2 + \beta_p \gamma_p - \frac{\alpha_p \beta_p \varepsilon_p^2}{\alpha_p + \beta_p G_p} \tag{4.29}$$

where, $\mathbf{M}_p \in R^{n_\theta \times n_\theta}, \forall_p \in N$ is a positive definite matrix that defines the shape and orientation of the ellipsoid; $\alpha_p \in [0, 1], \forall_p \in N$ is the forgetting factor for the old data and $\beta_p \in [0, 1], \forall_p \in N$ is the selecting factor for the new data. The OBE starts by assuming an ellipsoid that is big enough to contain $S^p, \forall_p \in N$ and a location for the center of the ellipsoid. As such, it is a often the case that the OBE algorithm is initialized with $\boldsymbol{\theta}_0 = \boldsymbol{\theta}_{(n_\theta \times 1)}$ and $\mathbf{M}_0 = \zeta \mathbf{I}_{(n_\theta \times n_\theta)}$. Where ζ is a large constant, e.g. 1e+5, and $\mathbf{I}_{(n_\theta \times n_\theta)}$ is the identity matrix. Assuming that $\mathbf{M}_{p-1}$ and σ_{p-1} are the positive definite matrix that is symmetric and a nonzero scalar at instant $p-1$, respectively, Eq. (4.30) signifies the corresponding ellipsoid [15].

$$\Omega(\hat{\boldsymbol{\theta}}_{p-1}, \mathbf{M}_{p-1}) = \{\boldsymbol{\theta} \in R^{n_\theta} | (\boldsymbol{\theta} - \hat{\boldsymbol{\theta}}_{p-1})^{\mathbf{T}} (\mathbf{M}_{p-1})^{-1} (\boldsymbol{\theta} - \hat{\boldsymbol{\theta}}_{p-1}) \leq \sigma_{p-1}^2\} \tag{4.30}$$

If a new data point (y^p, ψ_p) is given, the ellipsoid that encloses $\Omega(\hat{\boldsymbol{\theta}}_{p-1}, \mathbf{M}_{p-1}) \cap S^p$ is given by

$$\begin{aligned}\Omega(\hat{\boldsymbol{\theta}}_p, \mathbf{M}_p) = \{\boldsymbol{\theta} \in R^{n_\theta} | \sigma_p (\boldsymbol{\theta} - \hat{\boldsymbol{\theta}}_{p-1})^{\mathbf{T}} (\mathbf{M}_{p-1})^{-1} (\boldsymbol{\theta} - \hat{\boldsymbol{\theta}}_{p-1}) \\ + \beta_p (y_p - \psi_p^T \boldsymbol{\theta})^2 \leq \alpha_p \sigma_{p-1}^2 + \beta_p \gamma_p^2\}\end{aligned} \tag{4.31}$$

It is shown in [15] that, using (4.18)–(4.22) in (4.23) results in

$$\Omega(\hat{\boldsymbol{\theta}}_p, \mathbf{M}_p) = \{\boldsymbol{\theta} \in \Re^{n_\theta} | (\boldsymbol{\theta} - \hat{\boldsymbol{\theta}}_p)^{\mathbf{T}} (\mathbf{M}_p)^{-1} (\boldsymbol{\theta} - \hat{\boldsymbol{\theta}}_p) \leq \sigma_p^2\} \tag{4.32}$$

The values of α_p and in (4.25)–(4.29) and (4.23) rely on the criterion selected for OBE optimization. In the methods proposed by Fogel and Huang [13], the volume of the ellipsoid $V_{vol,p}(\alpha_p, \beta_p, \lambda_p)$ and sum of squares of semi-axes of the ellipsoid $V_{trace,p}(\alpha_p, \beta_p, \lambda_p)$ are minimized. For each optimization algorithm, the values of α_p and β_p are set equal to $1/\sigma_p^2$ and λ_p/γ_p^2, respectively. However, in the first degenerate minimum-volume algorithm which is proposed by Dasgupta and Huang [15] the minimized parameter is rather σ_p^2 . This algorithm is based on the premise that $\alpha_p = 1 - \lambda_p$ and $\beta_p = \lambda_p$.

$$V_{vol,p}(\alpha_p, \beta_p, \lambda_p) = det(\sigma_p^2 \mathbf{M}_p) \tag{4.33}$$

$$V_{trace,p}(\alpha_p, \beta_p, \lambda_p) = trace(\sigma_p^2 \mathbf{M}_p) \tag{4.34}$$

For each of the above-mentioned algorithms optimum value for λ_p is decided relying on the following procedures. In minimizing $V_{vol,p}(\alpha_p, \beta_p, \lambda_p)$, λ_p is obtained by solving the following quadratic equation with the coefficients as defined by in (4.28).

$$a_1\lambda_p^2 + a_2\lambda_p + a_3 \qquad (4.35)$$

$$\begin{cases} a_1 = (n_p - 1)\sigma_{p-1}^4 G_p^2 \\ a_2 = \{(2n_\theta - 1)\gamma_p^2 - \sigma_{p-1}^2 G_p\}\sigma_{p-1}^2 G_p \\ a_3 = \{n_\theta(\gamma_p^2 - \varepsilon_p^2) - \sigma_{p-1}^2 G_p\}\gamma_p^2 \end{cases} \qquad (4.36)$$

The optimum value for λ_p is given by:

$$\lambda_p = \begin{cases} 0 & if\ a_3 \geq 0. \\ \frac{-a_2 + \sqrt{a_2^2 - 4a_1 a_3}}{2a_2} & if\ a_3 < 0, \end{cases} \qquad (4.37)$$

In case of minimizing $V_{trace,p}(\alpha_p, \beta_p, \lambda_p)$, λ_p is the solution of a polynomial as given by Eq. (4.38).

$$\lambda_p^2 + b_1\lambda_p^2 + b_2\lambda_p + b_3 = 0 \qquad (4.38)$$

where, $b_1 = \frac{3\gamma_p^2}{\sigma_{p-1}^2 G_p}$

$$b_2 = \{\gamma_p^2 G_p[V_{trace,p-1}(\gamma_p^2 - \varepsilon_p^2) - \sigma_{p-1}^4 h_p] + 2\gamma_p^2[\gamma_p^2 G_p V_{trace,p-1} - \sigma_{p-1}^2 \gamma_p(\gamma_p^2 - \varepsilon_p^2)]\}/\zeta_p$$

$$b_3 = \gamma_p^4[(\gamma_p^2 - \varepsilon_p^2)V_{trace,p-1} - \sigma_{p-1}^4 \gamma_p]/(\sigma_{p-1}^2 \zeta_p)$$

$$h_p = \psi^T \mathbf{M}_{p-1}^2 \psi$$

$$\zeta_p = \sigma_{p-1}^4 G_p^2[G_p V_{trace,p-1} - \sigma_{p-1}^2 h_p]$$

Then, the optimum values for λ_p is calculated as

$$\lambda_p = \begin{cases} p & if\ b_3 \geq 0, \\ \lambda_p^* & otherwise \end{cases} \qquad (4.39)$$

where, λ_p^* is positive real root of (4.39).

Until now the calculation of model parameters is discussed. Equally important, is the estimation of prediction intervals. Following the discussion given in [8], for each estimated $\hat{\boldsymbol{\theta}}_p \in R^{n_\theta}$ and ellipsoid $\Omega(\hat{\boldsymbol{\theta}}_{p-1}, \mathbf{M}_p)$ an uncertainty interval $[\boldsymbol{\theta}_{min,p}(k), \boldsymbol{\theta}_{max,p}(k)]$ can be included by evaluating the following equations.

$$\boldsymbol{\theta}_{p,min}(k) = \min_{\theta \in \Omega_p} \boldsymbol{\theta}_p, k \in [1, n_\theta] \qquad (4.40)$$

$$\boldsymbol{\theta}_{p,max}(k) = \max_{\theta \in \Omega_p} \boldsymbol{\theta}_p, k \in [1, n_\theta] \qquad (4.41)$$

Accordingly, the parameter uncertainty bounds are estimated by (4.42) and (4.43).

$$\boldsymbol{\theta}_{p,min}(k) = \boldsymbol{\theta}_p(k) - \sigma_{p-1}\{\mathbf{M}_p(k,k)\}^{1/2} \qquad (4.42)$$

$$\boldsymbol{\theta}_{p,max}(k) = \boldsymbol{\theta}_p(k) - \sigma_{p-1}\{\mathbf{M}_p(k,k)\}^{1/2} \qquad (4.43)$$

It is easy to recognize at this stage that, for the OBE procedure the confidence interval or the model threshold relies on the size and orientation of the ellipsoid that contains the feasible set. Assuming that the optimum model is represented by $\psi_p^T \hat{\theta}$, the prediction interval may also be calculated based on Fig. 4.2 [9]. With CI_{BE} signifying the confidence interval for bounded error approach, the region that can be considered acceptable is described as:

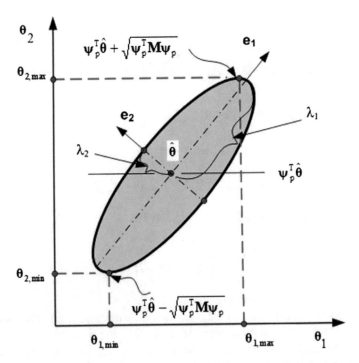

Fig. 4.2 Schematic diagram indicating the relationship between the size of minimum volume ellipsoid and confidence limits [9]

$$\psi_{\mathbf{p}}^{\mathbf{T}} - CI_{BE} - \gamma_p \le \psi_{\mathbf{p}}^{\mathbf{T}}\hat{\theta} - CI_{BE} + \gamma_p \qquad (4.44)$$

$$CI_{BE} = \sqrt{\psi_{\mathbf{p}}^{\mathbf{T}}\mathbf{M}\psi_p} \qquad (4.45)$$

4.2.2   Uncertainty Calculation for Nonlinear in Parameter Models

The neural network and TSK fuzzy models described in Chap. 3 are all nonlinear in parameter models. They can be rewritten in a generic form as (4.46).

$$y^{(p)} = f(\mathbf{u}; \theta^*) + \varepsilon^{(p)} \qquad (4.46)$$

Where, θ^* stands for vector of true values of the model parameters and ε is the model and measurement error or the error associated with $f(\mathbf{u}^{(p)}; \theta^*)$ in modeling the system. Approximations on the model parameters θ^* can be obtained by minimizing the sum of square of model errors function $V(\theta)$. If the cumulative sum of the errors is written as

$$\mathbf{V}(\theta) = \sum_{p=1}^{Nd} (y^p - f^p(\mathbf{u}; \theta))^2$$

Then,

$$\hat{\theta} = arg\,min \sum_{p=1}^{Nd} (y^p - f^p(\mathbf{u}; \theta))^2$$

And, the predicted value for a new data point $\mathbf{u}_0$ is

$$\hat{y}_0 = f(\mathbf{u}_0; \hat{\theta}) \qquad (4.47)$$

In general, it is possible to make approximation of the model parameters very close to the actual value. But, exact match is difficult attributed to many reasons, some of which are poor optimization algorithm, and factors not considered during modeling.

Linearization of the nonlinear model is needed in order to determine the model confidence interval. For linearization, Taylor series expansion to the first order of the model equation may be adopted. Besides, we may assume that the output is featured by random model error properties. Approximating (4.47) by first order Taylor series expansion

$$f(\mathbf{u}_0; \hat{\theta}) = f(\mathbf{u}_0; \theta^*) + (\boldsymbol{J}_0)^T (\hat{\theta} - \theta^*) \qquad (4.48)$$

where,

$$(\mathbf{J}_0)^T = \left[\frac{\partial f(\mathbf{u}; \boldsymbol{\theta}^*)}{\partial \theta_1^*} \quad \cdots \quad \frac{\partial f(\mathbf{u}; \boldsymbol{\theta}^*)}{\partial \theta_j^*} \quad \cdots \quad \frac{\partial f(\mathbf{u}; \boldsymbol{\theta}^*)}{\partial \theta_{nm}^*} \right]$$

The subscript 0 represents any set point other than that used for the estimation of the model parameters. The assumption that $\hat{\theta}$ and ε_0 are statistically independent in the estimation of model confidence intervals leads to the following variance relations. Using the Taylor series approximation (4.48), the modeling error is expressed as

$$y_0 - \hat{y}_0 \approx \varepsilon_0 - (\mathbf{J}_0)^T (\hat{\boldsymbol{\theta}} - \boldsymbol{\theta}^*) \tag{4.49}$$

From (4.49), expectation of the modeling error is given by

$$E\{y_0 - \hat{y}_0\} \approx E\{(\mathbf{J}_0)^T (\hat{\boldsymbol{\theta}} - \boldsymbol{\theta}^*)\}$$

Introducing the assumption that measurement noise at different points are uncorrelated,

$$Cov\{\varepsilon\} = E\{\varepsilon^2\} = \sigma^2 \mathbf{I}$$

And, covariance of the model parameters are approximated by

$$Cov\{\hat{\boldsymbol{\theta}}\} = E\{\hat{\boldsymbol{\theta}} - \boldsymbol{\theta}^*\} = \sigma^2 [(\mathbf{J}_0)^T (\mathbf{J}_0)]^{-1}$$

With the covariance matrix approximated, the confidence region is expressed as

$$(\hat{\boldsymbol{\theta}} - \boldsymbol{\theta})^T \mathbf{M}_c (\hat{\boldsymbol{\theta}} - \boldsymbol{\theta}) \leq (nm + 1) F_{\alpha, N_d - nm - 1}^{nm+1}$$

where, $\mathbf{M}_c = \sigma^2 [(\mathbf{J}_0)^T * (\mathbf{J}_0)]^{-1}$ and $\mathbf{J}(\hat{\boldsymbol{\theta}})$ is called the Jacobian matrix and it is given by

$$\mathbf{J}(\hat{\boldsymbol{\theta}}) = \begin{bmatrix} \frac{\partial f(\mathbf{u}^1; \boldsymbol{\theta}^*)}{\partial \theta_1} & \cdots & \frac{\partial f(\mathbf{u}^1; \boldsymbol{\theta}^*)}{\partial \theta_j} & \cdots & \frac{\partial f(\mathbf{u}^1; \boldsymbol{\theta}^*)}{\partial \theta_{nm}} \\ \vdots & \ddots & \vdots & \cdots & \vdots \\ \frac{\partial f(\mathbf{u}^p; \boldsymbol{\theta}^*)}{\partial \theta_1} & \cdots & \frac{\partial f(\mathbf{u}^p; \boldsymbol{\theta}^*)}{\partial \theta_j} & \cdots & \frac{\partial f(\mathbf{u}^p; \boldsymbol{\theta}^*)}{\partial \theta_{nm}} \\ \vdots & \vdots & \vdots & \ddots & \vdots \\ \frac{\partial f(\mathbf{u}^{Nd}; \boldsymbol{\theta}^*)}{\partial \theta_1} & \cdots & \frac{\partial f(\mathbf{u}^{Nd}; \boldsymbol{\theta}^*)}{\partial \theta_j} & \cdots & \frac{\partial f(\mathbf{u}^{Nd}; \boldsymbol{\theta}^*)}{\partial \theta_{nm}} \end{bmatrix}$$

Details of the derivatives are given in Appendix B. Now, using $\mathbf{J}(\hat{\boldsymbol{\theta}})$ and assuming a confidence level of $(1 - \alpha)$, the confidence region for a new prediction y_{new} is

$$CI \approx t_{a/2, N - nm} \{ \hat{\sigma}_{ref}^2 (1 + \mathbf{J}(\hat{\boldsymbol{\theta}})_{new}^T [(\hat{\boldsymbol{\theta}})_{ref}^T (\hat{\boldsymbol{\theta}})_{ref}]^{-1} (\hat{\boldsymbol{\theta}})_{new}^T) \}^{\frac{1}{2}} \tag{4.50}$$

where, $t_{a/2, N-nm}$ is the percentage value of $t-$ distribution that leaves a probability of $\alpha/2$ in the upper tail and $(1 - \alpha/2)$ in the lower tail. $(N - nm)$ is the degrees of freedom. Because σ^2 is unknown, the unbiased estimate of it, that is $\hat{\sigma}^2_{ref}$ is used in the calculation of CI.

$$\hat{\sigma}^2_{ref} = \frac{1}{N - n_\theta} \sum_{p=1}^{N} (y^p - f(\mathbf{u}; \hat{\boldsymbol{\theta}}))^2$$

Equation (4.50) is developed under the condition that the number of training samples goes to infinite, the modeling error is identically and independently distributed (iid), and the errors are normally distributed with zero mean and constant variance, $N(0, \sigma^2 I)$. The assumption of constant variance treats the output data set with the same noise variation, which is far from reality. In [16], a modification was suggested to improve the noise model. Instead of using Eq. (4.50) as it is, the value $\hat{\sigma}_{ref2}$ is replaced by S_a^2, which is calculated after clustering the input data and estimating unbiased variance for each cluster. Details of the relevant equations are:

$$\alpha_C(x) = exp\left(-\frac{1}{\sigma_c^2} \|\mathbf{u} - \mathbf{u}_C\|^2\right)$$

$$s_c^2(\mathbf{u}) = \frac{1}{n_c - n_u} \|y_c - f(\mathbf{u}, \boldsymbol{\theta})\|^2$$

$$s_a^2(\mathbf{u}) = \frac{\sum\limits_{c=1}^{C} \alpha_c(\mathbf{u}) s_c^2}{\sum\limits_{c=1}^{C} \alpha_c(\mathbf{u})}$$

$$CI \approx t_{a/2, N-nm} \{s_a^2 (1 + J(\hat{\boldsymbol{\theta}})_{new}^T [(\hat{\boldsymbol{\theta}})_{ref}^T (\hat{\boldsymbol{\theta}})_{ref}]^{-1} (\hat{\boldsymbol{\theta}})_{new}^T) \}^{\frac{1}{2}} \qquad (4.51)$$

where, S_c^p is the covariance of the model error in cluster c ; $\mathbf{r}_c$ is the data set used for training the model in the region defined by cluster c ; y_c is the target corresponding to the input $\mathbf{u}_c$; and n_c is the number of data points in cluster c. This formulation of variance has been adopted by Muoz and Sanz-Bobi [17] in formulating a probabilistic neural network model.

To deal with the unrealistic nature of the assumption on the measurement errors, two additional modifications on (4.50) were presented by Shao et al. [18]. The first approach is based on the assumption that the model error is featured by $N(0.|y_{0,p} - \hat{y}_{0,p}|)$. Accordingly, (4.50) is modified based on the ratio between the prediction error and the standard deviation of the whole error vector. That is,

$$CI_p \approx \lambda . t_{a/2, N-nm} \{\hat{\sigma}^2_{ref} (1 + J(\hat{\boldsymbol{\theta}})_{new}^T [(\hat{\boldsymbol{\theta}})_{ref}^T (\hat{\boldsymbol{\theta}})_{ref}]^{-1} (\hat{\boldsymbol{\theta}})_{new}^T) \}^{\frac{1}{2}} \qquad (4.52)$$

where,

$$\lambda_p = \frac{|y_{0,p} - \hat{y}_{0,p}|}{std|y_{0,p} - \hat{y}_{0,p}|}, \quad p = 1, 2, \ldots, N$$

The second approach proposed in [18] is based on the hypothesis that confidence intervals are also dependent on the density of the data. The resulting equation is given by

$$CI_s = \frac{2}{1 + \rho/\rho_{max}} CI \tag{4.53}$$

where, ρ is the density estimation, ρ_{max} is the maximum value of ρ ; CI is the confidence interval as calculated from (4.50).

4.3   Fault Detection

As defined in Chap. 2 and Sect. 2.1, fault detection is the process of knowing the presence of a fault in a system and time of detection. It may be realized by comparing the actual signal with a reference signal or based on the comparison of residuals and model uncertainties calculated relying on pre-defined error assumptions. The reference model for the raw use or residual calculation can be formulated applying historical data based techniques (Chap. 3, Sect. 3.2) or through the use of models developed from first principles. The critical decision in fault detection system design is whether to use fixed model threshold or adaptive threshold, Fig. 4.3. For a system whose operating region is not so much varying and in a condition where the adopted modeling technique is difficult to be described by an adaptive model uncertainty

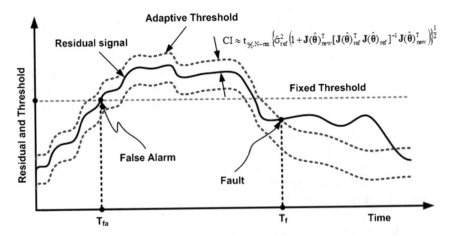

Fig. 4.3 Concepts of fixed threshold and adaptive threshold

equation (e.g. auto-associative neural networks), fixed threshold techniques may be considered effective. In this case three-sigma approach is the easiest to apply. In practice, for such systems as GTG, HRSG and SAC, the operating region varies with external conditions and the load demand. Hence, using constant model threshold may lead to false alarms. For the three systems, the adaptive threshold technique is indeed the perfect choice. In this thesis, the confidence interval equation as defined by (4.50) is employed for designing the fault detetion and diagnosis system. For controlling sensitivity of the adaptive threshold to faults, the α-cut in the CI equation could be manipulated.

4.4 Fault Diagnostics

Fault diagnosis is the task of identifying the location and time of occurrence of a fault based on the available signals that can be from analytical residual calculation, hardware redundancy, or heuristic information. The diagnosis system must be able to map the diagnostic signal to defined state of the system, which includes the faulty state and normal state.

Fault diagnosis can be conceived as the mapping of the diagnostic signals to the fault states. That is,

$$\mathbf{V} : \mathbf{S} \rightarrow \mathbf{F}$$

where, $\mathbf{S} R^{n_s}$: the diagnostic signal; $\mathbf{F} \in R^{n_f}$: system fault; n_s : number of diagnostic signals; n_f : number of faults. The mapping function can be designed to operate on binary diagnostic signals, multi-valued diagnostic signals or continuous diagnostic signals [19]. The following section will consider the design of a diagnosis system based on the assumption that the available signals are binary diagnostic signals.

4.4.1 Mapping Binary Diagnostic Signals into the Space of Faults

The signals available for fault diagnosis could be sourced by residual calculation through analytical techniques, simple limit checking, hardware redundancy or heuristic methods. In binary fault diagnostics techniques, binary forms of the signals $S_j \in \{0, 1\}(j = 1, \ldots, n_s)$ are first formulated by two value evaluation of the raw diagnostic signals. Magnitude of a residual beyond the NOC model threshold is assigned a value of 1 while insignificant changes take a value of zero. The model responsible for fault diagnosis then realizes the mapping

$$\mathbf{V} : \mathbf{S} \in \{0, 1\}_{ns} \rightarrow \mathbf{F} \in \{0, 1\}_{n_f}$$

Fig. 4.4 Binary diagnostics matrix (BDM) for a generic system

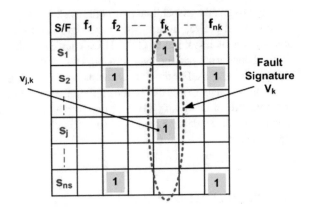

The sets of signals $\mathbf{F} = \{s_j : j = 1, \ldots, n_s\}$ and the sets of faults $\mathbf{F} = \{f_k : k = 1, \ldots, n_k$ are often presented as binary trees, ifthen rules (4.54), logic functions (4.55), or binary diagnostic matrix (Fig. 4.4). It is possible to define the first three using the binary diagnostic matrix. In the if...then design, the k-th rule is written as

$$R_k : if\ s_1\ is\ v_{1,k}\ and\ s_2\ is\ v_{2,k}\ and\ \ldots s_j\ is\ v_{j,k} \ldots and..s_{ns}$$
$$is\ v_{ns,k}\ then\ fault\ f_k \tag{4.54}$$

where, $v_{j,k}$ in the rule takes a value of either 0 or 1 if the design of the diagnostic system is based on binary evaluation of the residuals. The rule given by (4.54) corresponds to the k-th column in the binary diagnostic matrix. A simplified logic function $\alpha(f_k)$ that signifies a rule from (4.54) can be written as

$$\alpha(f_s) = s_1 \wedge s_2 \wedge \ldots \wedge s_{ns} \tag{4.55}$$

Once the BDM is designed, the mapping from binary signals to fault sets can easily be described by a Cartesian product relation between $\mathbf{S}$ and $\mathbf{F}$.

$$\mathbf{V} = \mathbf{S} \times \mathbf{F} \tag{4.56}$$

In Eq. (4.56), a value of $V_{j,k}=1$ is equivalent to saying that the existence of the fault f_k is partly featured by changes in the signal s_j. A value of $v_{j,k}$ equal to 0 has the meaning of no relation between f_k and s_j. In the fault signature formulation, the fault signature for fault f_k is expressed as $v_k = [v_{1,k} \quad v_{2,k} \ldots v_{ns,k}]^T$. If v_k is extended to the whole fault set $\mathbf{F} = \{f_k : k = 1, \ldots, nk\}$, it results in

$$\mathbf{V} = \begin{bmatrix} v_{1,1} & v_{1,2} & \cdots & v_{1,nk} \\ v_{2,1} & v_{2,2} & \cdots & v_{2,nk} \\ \vdots & \vdots & \ddots & \vdots \\ v_{ns,1} & v_{ns,2} & \cdots & v_{ns,nk} \end{bmatrix}$$

These concepts will be used in the design of NF based fault diagnostic systems. One advantage of the binary diagnostic approach is simplicity. However, it has the drawback of overlooking the uncertainty in the diagnostic system. The next section extends the binary approach to a design based on fuzzy systems.

4.4.2 Diagnostics Using Fuzzy Systems

Fuzzy systems are used in fault diagnosis design either as fault classification tool or for formulating an automatic inference system. In cases where there is no predefine knowledge about fault-symptom causalities, classification methods are applied to map the diagnostic signals to the space of faults. Among the common classification techniques is fuzzy clustering. In realizing fault diagnosis through fuzzy clustering, first multiple clusters are formed and for a given diagnostic signal fuzzy membership value of the signal for all the existing classes are estimated. Based on the highest membership value, the feature vector is classified to a specific class hence diagnosis. This technique has been applied in the research work of Yiping, Yiping et al. [20], Zogg, Shafai et al. [21], Babbar, Ortiz et al. [22].

Excellent features of fuzzy systems are: they possess fuzzy operators (see Appendix A) capable of dealing with imprecise information, and they have the characteristic that they use ifthen inference system as part of their building block. The two features allow the implementation of fuzzy systems for the design of fault diagnostics method for a system whose fault-symptom causalities are known a prior. Causal relationships between faults and diagnostics signals can be generated from a binary diagnostic matrix or based on experts knowledge. The application of fuzzy if then reasoning method combined with other analytical symptoms generation based fault detection techniques can be found in the works of Zhao and Upadhyaya [23], Zio and Gola [24], and Razavi-Far et al. [25].

The fault diagnosis system design developed in this thesis follows the general reasoning structure of a fuzzy rule based system. First, linguistic variables $v_j = \{v_{j,1}, v_{j,2}, \ldots, v_{j,nj}\}$ are attributed to each residual r_j or its magnitude $\|r_j\|$ and the residuals are fuzzified. Then, the fuzzy diagnostic signals are used in the fuzzy diagnostic inference engine to infer on none exact facts about the possible fault. General structure of the fuzzy diagnostic system is shown in Fig. 4.4. It is characterized by fuzzification and inference block only. The inference block has inherently sets of rules designed from binary diagnostic matrix or from knowledge of experts.

4.4.3 Rules of Inference and Fault Diagnostics

The fuzzy diagnostic system works based on sets of rules defined on the grounds of the relationship between diagnostic signals and fault sets. One way for developing the rules is to use two value decomposition of the diagnostic signal and formulate a

Binary Diagnostic Matrix (BMD). From there, sets of rules can be arranged separating or grouping columns of the matrix possessing similar premises as independent state. Refer to (4.54) for the fuzzy rule in a generic design.

For fault free state, the signal values $v_{i,j}(j-1,\ldots,n_s)$ are all zero. The design based on binary evaluation eliminates the fact that diagnostics signals could posses some bias in their characteristics. To this end, an alternative design approach is to use multi-valued evaluation that could be relying either on crisp features or imprecise signals. The crisp consideration is just a slight modification of the binary approach for the vague information not yet included. In fuzzy evaluation of a diagnostic signal, any number of linguistic variables $v_j = \{v_{j,1}, v_{j,2}, \ldots, v_{j,nj}\}$ can be assumed leading to multi-valued evaluation of the diagnostic signal.

The third way to formulate the rules of the inference engine is to use experts knowledge. The drawback to this approach, though, is the design highly relies on the experience of the expert and it is frequently difficult to come up with enough sets of rules. It can be considered as an advantage if it is used to supplement the other two methods. In the diagnostic system design considered in this thesis, the rules are fuzzy based but instead of designing them directly from input-output data using fuzzy Mamdani model [26], they are derived from the BDM.

In fuzzy diagnostic system (Ref. to Fig. 4.5), the outputs from the inference engine are certainty factors $\mu_{ST,k}(k = 1, \ldots, n_k)$ for different states: states with complete efficiency or normal operation (ST_0), and faulty state ST_k. In the rules generated for binary diagnostic signals, the antecedents are in the form of conjunction. If the rules are to be adopted for fuzzy diagnostic signals, then each of the entries in the BDM should be linked with a corresponding fuzzy set that matches with the linguistic variables assigned for the diagnostic signal. Figure 4.6 indicates the base BDM, the Fuzzy Diagnostic Matrix (FDM) and the corresponding fuzzy set for a system having two rules and two diagnostic signals.

Applying the FDM developed from BDM, the rules from (4.54) are modified and expressed as

$$R_k : if\ s_1\ is\ Z\ and\ s_2\ is\ NZ\ and\ \ldots s_j\ is\ Z\ \ldots\ and\ldots s_{ns}$$
$$is\ Z\ then\ fault\ f_k \tag{4.57}$$

where, Z and NZ stand for zero and nonzero, respectively. The corresponding generic FDM is given in Fig. 4.7. For j-th diagnostic signal in the k-th rule, the degree of activation of the signal in the k-th rule has a linguistic value

$$\mu(f_k, s_j)_{FDM} = \begin{cases} \mu_{Z,J} & for\ c_{k,j} = Z \\ \mu_{NZ,J} & for\ c_{k,j} = NZ \end{cases}$$

The activation level of each of the rules in (4.57) is calculated applying fuzzy operators, Appendix A. For the $k-th$ rule, it appears as

$$\mu(ST_k)_{FDM} = \mu(ST_k, s_1) \otimes \mu(ST_k, s_2) \otimes \cdots \otimes \mu(ST_k, ns)$$

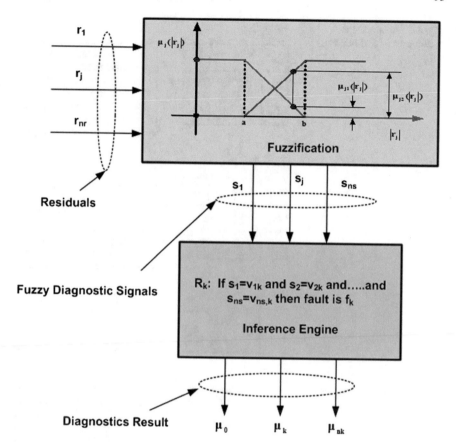

Fig. 4.5 Fuzzy fault diagnostic system

where, $\otimes$ is the fuzzy conjunction operator. The conjunction operators are realized by t-norm operators among which are PROD operator, MIN operator and MEAN operator. The MIN operator considers the minimum or lowest value from the rule premise while MEAN takes on the average.

$$\mu(ST_k)_{PROD} = \prod_{j=1}^{ns} \mu(ST_k, s_j) \qquad (4.58)$$

$$\mu(ST_k)_{MIN} = inf\{\mu(ST_k, s_1, \mu(ST_k, s_2,), \ldots, \mu(ST_k, s_{ns}\} \qquad (4.59)$$

$$\mu(ST_k)_{MEAN} = \frac{1}{ns} \sum_{j=1}^{ns} \mu(ST_k, s_j) \qquad (4.60)$$

Fig. 4.6 Relationships between BDM and FDM

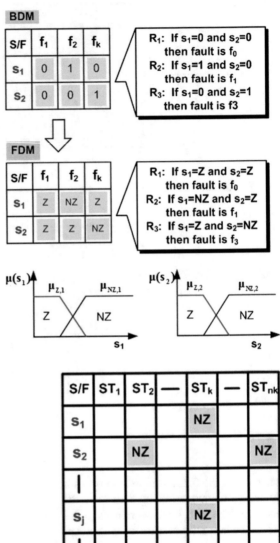

Fig. 4.7 Fuzzy diagnostics matrix for a generic system

S/F	ST_1	ST_2	—	ST_k	—	ST_{nk}
s_1				NZ		
s_2		NZ				NZ
|						
s_j				NZ		
|						
s_{ns}		NZ				NZ

After the rule activation values are calculated by anyone of these equations selected for the whole set of rules, a rule with the highest activation level or certainty factor will be considered as indicating the corresponding fault highly likely to appear. The certainty factor together with the state identification character defines the output from

the fuzzy diagnostic system. The diagnostic result for state ST_k may be expressed as ordered pairs

$$D_{ST_k} = \{\langle ST_k, \mu_{ST_k} : \mu_{ST_k} > 0\}$$

4.5 Summary

First aim of this chapter has been on the discussion of the equations behind the estimation of model uncertainity. The second target was, to put address the design of a robust fault detection and diagnosis system based on adaptive confidence intervals. The main ideas are summaried as follows:

- In Sect. 4.2, model confidence interval equations are formulated for linear model and nonlinear models, respectively. It was shown that confidence interval for a linear model can be formulated assuming iid errors or bounded errors. For the bounded error assumption, we have used OBE algorithm to estimate model parameters and quantify the uncertainty. For the nonlinear case, while keeping the iid assumption, the confidence intervals are constructed applying Taylor series expansion that led to a Jacobian matrix. In the same section, a discussion was made on modified forms of confidence interval equations as suggested by other authors.
- In Sect. 4.3, the fault detector is covered. The difference between a constant threshold and adaptive threshold from the point of view of robust fault detector is elaborated.
- In Sect. 4.4, the design of a fault diagnosis system based on binary or fuzzy evaluation of residuals is elucidated. In the fuzzy design, a method that assumes two value evaluations is adopted.

Finally, it can be said that the current chapter has addressed the adaptive model uncertaininty equations in a way suitable for fault detection and diagnosis. The integration of the adaptive CI model with the fuzzy diagnosis steps is main contribuition of the chapter. In the next Chapter, the equations will be used to formulate a fault detection and diagnosis system that takes into account integrated operation of multiple systems.

References

1. Weyer, E., Szederknyi, G., & Hangos, K. (2000). Grey box fault detection of heat exchangers. *Control Engineering Practice, 8*, 121–131.
2. Hill, D. J., & Minsker, B. S. (2009). Anomaly detection in streaming environmental sensor data: A data-driven modeling approach. *Environmental Modelling and Software, 25*, 1014–1022.
3. Mencar, C., Castellano, G., & Fanelli, A. M. (2005). Deriving prediction intervals for neuro-fuzzy networks. *Mathematical and Computer Modelling, 42*, 719–726.

4. Mrugalski, M. Korbicz, J. (2007). Least mean square vs. Outer bounding ellipsoid algorithm in confidence estimation of the GMDH neural networks. *Proceedings of the 8th International Conference on Adaptive and Natural Computing Algorithms, Part II Warsaw*, Poland: Springer.

5. Sderstrm, T., & Stoica, P. (1988). *System Identification.*, Prentice Hall international series in systems and control engineering Englewood Cliffs: Prentice-Hall, Inc.,

6. Ljung, L. (1999). *System Identification: Theory for the User.*, Prentice Hall Information Series Englewood Cliffs: Prentice Hall.

7. Witczak, M., Korbicz, J., Mrugalski, M., & Patton, R. J. (2006). A GMDH neural network-based approach to robust fault diagnosis: Application to the DAMADICS benchmark problem. *Control Engineering Practice, 14*, 671–683.

8. Milanese, M., Norton, J., Piet-Lahanier, H., & Walter, E. (1996). *Bounding Approches to System Identification*. New York: Plenum Press.

9. Walter, E., & Pronzato, L. (1997). *Identification of Parameteric Models from Experimental Data*. Berlin: Springer.

10. Kelnhofer, R. W. (1997). *Applications of Interval Methods to Parameter Set Estimation from Bounded-Error Data*. United Staes - Wisconsin: Marquette University.

11. Deller, J. R., Nayeri, J. M., & Liu, M. S. (1994). Unifying the landmark developments in optimal bounding ellipsoid identification. *International Journal of Adaptive Control and Signal Processing, 8*, 43–60.

12. Arruda, L. V. R., Da Silva, I. N., & Amaral, W. C. (1996). A neuro-fuzzy method to parametric estimation with unknown-but-bounded-error. *In Systems, Man, and Cybernetics, IEEE International Conference on, 1*, 351–355.

13. Fogel, E., & Huang, Y. F. (1982). On the Value of Information in System Identification-Bounded Noise Case. *Automatica, 18*, 229238.

14. Shchweppe, F. (1968). Recursive state estimation: Unknown but bounded errors and system inputs. *Automatic Control, IEEE Transactions on, 13*, 22–28.

15. Dasgupta, S., & Yih-Fang, H. (1987). Asymptotically convergent modified recursive least squares with data-dependent updating and forgetting factor for systems with bounded noise. *IEEE Transaction on Information Theory, 33*, 383–392.

16. Da Silva Neves, C. A. M., Roisenberg, M., & Neto, G. S. (2009). A method to estimate prediction intervals for artificial neural networks that is sensitive to the noise distribution in the outputs. *Neural Networks, IJCNN 2009. International Joint Conference on*, 2238–2242.

17. Muoz, A., & Sanz-Bobi, M. A. (1998). An incipient fault detection system based on the probabilistic radial basis function network: Application to the diagnosis of the condenser of a coal power plant. *Neurocomputing, 23*, 177–194.

18. Shao, R., Zhang, J., Martin, E. B., & Morris, A. J. (1997). Novel approaches to confidence bound generation for neural network representations. *IEEE: Artifical Neural Networks*.

19. Korbicz, L., Koscielny, J. M., & Kowalczuk, Z. (2004). *Fault Diagnosis: Models, Aartificial Intelligence, Applications*. Berlin: Springer.

20. Yiping, L., Yi, S., & Zhiyan, L. (2000). An approach to fault diagnosis for non-linear system based on fuzzy cluster analysis. *Instrumentation and Measurement Technology Conference, IMTC 2000. Proceedings of the 17th IEEE, 3*, 1469–1473.

21. Zogg, D., Shafai, E., & Geering, H. P. (2006). Fault diagnosis for heat pumps with parameter identification and clustering. *Control Engineering Practice, 14*, 1435–1444.

22. Babbar, A., Ortiz, E. M., & Syrmos, V. L. (2009). Control and Automation, MED '09. 17th Mediterranean Conference on. *Fuzzy clustering based fault diagnosis for aircraft engine health management* (pp. 199–204).

23. Zhao, K., & Upadhyaya, B. R. (2005). Adaptive fuzzy inference causal graph approach to fault detection and isolation of field devices in nuclear power plants. *Progress in Nuclear Energy, 46*, 226–240.

24. Zio, E., & Gola, G. (2006). Neuro-fuzzy pattern classification for fault diagnosis in nuclear components. *Annals of Nuclear Energy, 33*, 415–426.
25. Razavi-Far, R., Davilu, H., Palade, V., & Lucas, C. (2009). Model-based fault detection and isolation of a steam generator using neuro-fuzzy networks. *Neurocomputing, 72*, 2939–2951.
26. Nelles, O. (2001). *Nonlinear System Identification: From Classical Approaches to Neural Networks and Fuzzy Models*. Berlin: Springer.

Chapter 5
Intelligent Fault Detection and Diagnostics

5.1 Introduction

This chapter contains the last part of the research methodology. On the bases of the methods discussed in Chaps. 3 and 4, it develops the planned FDD system. In Sect. 5.2, details of the proposed Intelligent Fault Detection and Diagnosis (IFDD) system are presented while Sect. 5.3 outlines the methods designed to deal with multiple subsystems. Section 5.4 of the chapter is dedicated to the presentation of the method proposed to address the effect of scheduled and forced changes on the NOC model construction. The steps in the design of OBF based NF models are elucidated in Sect. 5.5. In Sect. 5.6, the training of nonlinear models by a recursive method is explained. Section 5.7 gives the summary of the chapter.

5.2 Structure of the Proposed IFDD Framework

The proposed IFDD is defined by three serially connected modules: - Data Pre-processing Module (DPM), Fault Detection Module (FDEM), and Fault Diagnosis Module (FDIM). Figure 5.1 shows the overall structure including the input and output data. The inputs to the IFDD are the measured input data $u_j (j = 1, 2, \ldots, n_u)$, output data $y_q (q = 1, 2, \ldots, n_y)$ and data $h_i (h = 1, 2, \ldots, n_h)$ from other sources. The raw input-output data are pre-processed in the DPM and the resulting detection signals are passed to the FDEM where residuals $\mathbf{r} = [\mathbf{r}_u \, \mathbf{r}_y \, \mathbf{r}_s \, \mathbf{r}_h]^T$ are calculated and compared with their respective thresholds. Once out of bound residuals are detected, the FDIM is triggered and all the estimated residuals are conveyed to this module. The FDIM classifies the residual vector to the closest fault available in the FDIM knowledge library. The output from FDIM is given in the form of $< ST_j, \mu_j >$, for $j = 1, 2, \ldots, n_s$. ST_j is the jth state name.

© Springer International Publishing AG 2018
T.A. Lemma, *A Hybrid Approach for Power Plant Fault Diagnostics*,
Studies in Computational Intelligence 743,
https://doi.org/10.1007/978-3-319-71871-2_5

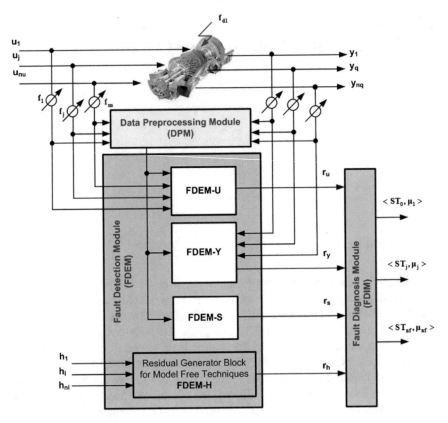

Fig. 5.1 Overall structure of IFDD

5.2.1   Data Pre-processing Module (DPM)

The DPM is included to deal data size, measurement noise, outliers, and figure size differences in the input-output data. In the design elaborated hereunder, the DPM is realized through the use of Discrete Wavelet Transfer (DWT), data normalization tools, PCA and NLPCA. Detail structure of the DPM is indicated in Fig. 5.2. The DWT is applied to separate the average signal from the noise signal so that the former could be used in the following modules. Mother wavelet function of Daubechies type with different orders is assumed to remove the noise as per user preference. Refer to Appendix C for the basic equations behind DWT. In our proposed structure, we have used DWT function from Wavelet Toolbox available in Matlab.

The second key element of DPM is the data normalization tool. Some of the data in the gas turbine are in the range of 0–1(e.g. VIGV position) while others in the range of 300–1000 (e.g. compressor discharge pressure). Similar cases exist in the HRSG and SAC. Using the signals avoiding pre-processing could lead to one

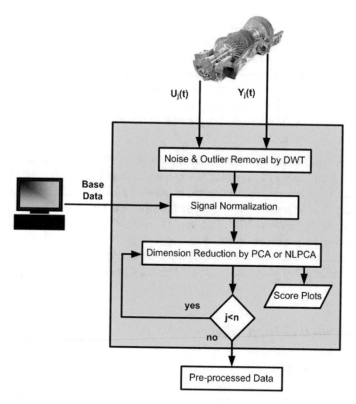

Fig. 5.2 Overall structure of IFDD

figure dominating the other during NF model training. In the proposed approach, the following techniques are considered in relation to data normalization.

- Normalizing the data using the maximum recorded value or against the design point data.
- Normalizing the data between 0 and 1.
- Normalizing the data in such a ways that mean of errors equal to zero and standard deviation equals to 1. The relevant equations and the corresponding flow chart are portrayed in Appendix C.

The third element that forms the DPM is the dimension reduction tool. The use of a dimension reduction helps to avoid long training times and curse of dimensionality. The DPM is equipped with PCA and NLPCA algorithms for the same reason. It can be shown that PCA and NLPCA have dimension reduction capabilities.

The NLPCA is based on the application of AANN [1]. It has better dimension reduction ability [2] than PCA. It was also proved suitable to handle nonlinear correlations between measured variables. The implementation of the two tools in the DPM follows the following steps.

- Check if the dimension of the input data n_u is one or two.
- If n_u is greater than two, apply PCA and reduce the dimension to two. In doing so, it is preferred if the variance explained by the remaining data is higher than 95%.
- If the variance stated in step-2 is lower than 95%, then the two retained principal components may not be enough to accurately represent the actual output. In that case NLPCA is applied. A dimension of two can be obtained from NLPCA assuming two nodes in the bottle neck layer of the auto-associative neural network (see Appendix C.2) model.

The proposed technique using PCA and NLPCA is different from other approaches in that instead of using PCA or NLPCA as a fault detection and diagnosis tool, they are used as data dimension reduction techniques only. Each has the characteristics of eliminating the portion of a data in the direction of low variance. As such, PCA and NLPCA when they are used in series connection with DWT will have the effect of further removing measurement noise hence supporting DWT. In fact PCA and NLPCA have difficulties in handling fault diagnosis, especially in handling multiple fault scenarios. Hence, the aforesaid proposed approach seems alternative application.

5.2.2   Fault Detection Module (FDEM)

The FDEM follows the data processing module. As indicated in Fig. 5.1, FDEM is comprised of four sub-modules for fault detection: FDEM-U for input sensors, FDEM-Y for fault detection based on output sensors, FDEM-S for fault detection through output soft sensors, and FDEM-H for fault detection relying on heuristic signals. The FDEM is featured by at least a number of NOC models as there are inputs and outputs. At times it is difficult to diagnose all faults from the available measurement alone. In the FDEM-S, NOC models developed applying first principles are used for fault detection. For instance, in the gas turbine quantities like temperature at the compressor outlet, efficiency of compressor, temperature at the turbine inlet, temperature at the turbine exit, turbine efficiency and pressure at the turbine inlet are very essential for health monitoring. Unless models of these parameters are formulated based on the same set of measured inputs, it will be difficult to reach to a reasonable diagnosis conclusion.

The sub-module design must not be confused with the classical observer designs in that linear state space approaches are mostly applied. The structure resembles the classical observer. Nevertheless, in the proposed approach the models in the FDEM-U, FDEM-Y and FDEM-S are designed based on NF approach that takes into account the effect of scheduled and forced operating regions, Sect. 5.4. Detail of the NF based fault detector in the FDEM is depicted in Fig. 5.3. The FDEM differs from other approaches in the following ways:

- To deal with the time varying residual signal (r) generated in the FDEM blocks, adaptive threshold technique based on IID or bounded error assumption is introduced.

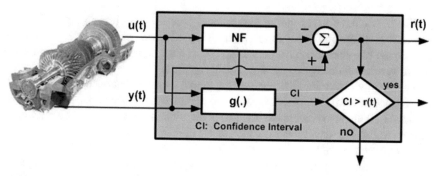

Fig. 5.3 Details in the FDEM-U or FDEM-Y or FDEM-S

- For the dynamic case, the local linear models in the consequent part of the dynamic NF model are constructed in the framework of Generalized Orthonormal Basis Functions (GOBFs), Sect. 3.2.2. The use of GOBF in the ARX model is well developed while the locally linear extension to nonlinear dynamic model is still at development stage. Implementation procedures behind the OBF based nonlinear models are elaborated in Sect. 5.6.

The training of the NF model in any of the sub modules of FDEM is performed by the methods discussed in Sect. 3.5. In estimating parameters relevant to the calculation of model confidence intervals, equations from Chap. 4 and Sect. 4.2 are employed. A detection of a fault is realized by comparing a residual with the corresponding confidence interval, Sect. 4.3. Fault detection is confirmed if one of the residuals is outside the estimated confidence interval.

Application Steps in the FDEM

The concepts considered hereunder include the calculation performed during model training and model application for fault detection. The models are developed by NF approach except for those in the FDEM-H. The following procedure is adopted for model training.

- Identify input and output data for the models in each operating region, M_j.
- Assume model optimization parameters for the learning algorithm. For the back propagation learning algorithm, these may include: learning rate (η), momentum term (α), maximum error tolerance (ε), and maximum number of iterations (N_i). The last two parameters are also applicable to other learning algorithms.
- Assume starting matrices for model parameters: centre matrix ($\mathbf{C}$), spread terms matrix (Σ) and matrix of consequent parameters ($\boldsymbol{\theta}$).
- Perform the model training until the allowed error or numbers of iteration mentioned in step-2 are achieved.
- Validate the model by a different set of data.

The flow chart corresponding to the stated procedure is given in Appendix D.2, Fig. D.2. Note that, in order to get preliminary estimates of the model parameters techniques like data clustering combined with least squares, particle swarm optimization or evolutionary computing may be applied.

During model training, it is not only the model matrices that need to be estimated. Confidence interval related matrix, $\mathbf{J_{ref}}$, and unbiased reference variance of the NOC model ($\hat{\sigma}_{ref}$) are also calculated. The above mentioned steps are applicable for each model required for developing the FDEM-U, FDEM-Y and FDEM-S.

In the execution of fault detection module, the NOC models are used to predict the output for a given input. To obtain the residuals, the predicted outputs are compared with an output that is either from the actual system or a reliable first principle model. As a main step to fault detection, the residuals are evaluated against the corresponding adaptive threshold. Any consecutive exceeding of calculated threshold value by the residual signal is considered as an indication of the presence of a fault in the system. It is worth mentioning that careful design of the model threshold is very important to avoid false alarms in case of highly sensitive design and missed faults in case of loosely set model threshold. In the proposed design 95% confidence level is assumed. However, the algorithm is arranged in such a way that setting of a different level of confidence is possible.

5.2.3 Fault Diagnosis Module (FDIM)

The FDIM takes residual signals and adaptive model threshold values from FDEM and provides diagnostic information in terms of state name and fault activation level. As stated in Chap. 4 and Sect. 4.4, the fault diagnosis system may be designed to work by mapping the residuals to binary diagnostic signals. Nevertheless, fuzzy systems are proposed for the binary approaches overlook diagnostic system uncertainty.

In the FDIM, binary evaluations of the residuals are first performed to formulate the BDM. Then, fuzzy rules are designed according to fault-symptom relations in the BDM. The fuzzy rules finally define the main diagnosis model. The detail steps are as follows:

- For a known fault type, generate residuals using the models in the FDEM.
- Perform binary evaluation on each residual applying the corresponding model confidence intervals, $CI_j(j = 1, 2, \ldots, n_s)$. Because it is a binary evaluation, magnitude of a residual, $\|r_j\|$, rather than the residual itself need to be considered.
- Formulate a Binary Diagnostic Matrix (BDM) from the results in step-2. A generic example is given in Chap. 4 (Table 4.1).
- Create diagnosis rules from the binary diagnostic matrix. The rules may incorporate the experts knowledge as well as knowledge from other sources.
- Translate the rules into Fuzzy Diagnostic Matrix (FDM).

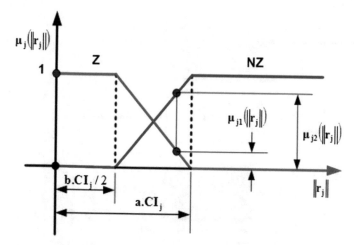

Fig. 5.4 Membership functions for residual evaluation

During application, the residuals are fuzzified by their respective membership functions. Figure 5.4 indicates forms of the membership functions adopted in the FDIM. Unique feature of the FDIM system is that, the confidence intervals (CI_j)- also called model threshold evolve with the dynamics of the system. That is to say supports (Ref. Appendix A) of the two membership functions, $\mu_{j,1}(\|r_j\|)$ and $\mu_{j,2}(\|r_j\|)$, are not fixed values. In defining the membership functions, the choices on $a \in [0.5, 1.5]$ and $b \in [0, 1]$ decides their location. Typical values that will be used in the FDIM are a equals to 1 and b equals to 1. Apparently, the decision on the two values control sensitivity of the fault diagnosis module.

5.3   Overall Plant Level Fault Detection and Diagnosis System

A cogeneration and cooling plant is defined by a primary system (in our case gas turbine), a Heat Recovery Steam Generator (HRSG) and a Steam Absorption Chiller (SAC). Considering the gas turbine alone, the health condition may be inferred from gas path related signals or considering signals in the auxiliary systems. The signals in the lube system and gas fuel system are typical examples. In fact, when the GTG is in temperature and load control, it runs in synchronous with the HRSG. We have witnessed an incidence where a problem in the diverter damper response time led lead to GTG trip. The steam absorption chiller runs by the steam from the steam header, with the source of the steam being from the HRSG. While it is an engineering practice to have the steam header to damp oscillating pressures as well as allow multiple steam sources for the SAC, the actual run is featured by continuous steam supply from the

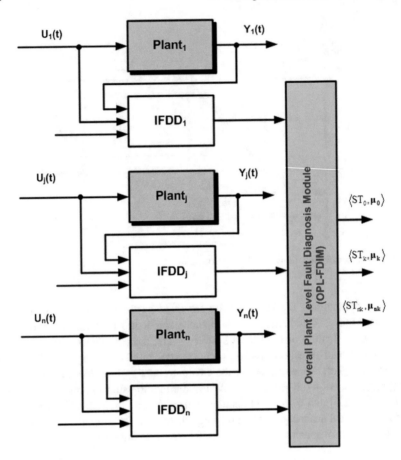

Fig. 5.5 Overall plant level intelligent fault diagnosis module (OPL-FDIM)

HRSG. Using these facts and taking into account the possibilities of future extension of the fault detection and diagnosis system to accommodate exergy based techniques, it is a wiser step if a fault detection and diagnosis frame work is formulated for the overall system.

In the IFDD, the fault detection for the overall system is realized by designing separate fault detection systems for the sub-systems and latter connect the outputs from each system to a common fault diagnosis block, Fig. 5.5. Details of the sub-system level detection systems are identical to the IFDD designed in Sect. 5.2. Important feature of the proposed approach is that, it is generic and the numbers of sub-systems are not limited. Besides, fuzzy neural networks can be used in the fault diagnosis block that could provide the output in the form of Sate numbers (ST_k) and fault certainty level (μ_k) which are hoped understandable by the plant operators.

5.4 Techniques to Accommodate Effects of Scheduled and Forced Operating Conditions on NOC Models

The design of a fault detection and diagnosis system needs to answer the questions that come with variable operating regions. A gas turbine may be designed to operate at two different regions. Including the start-up and shut down, a total of four major operating regions are possible. Figure 5.6 indicates a generalized case where there are N_{ap} number of operating regions with the transition regions being considered as fuzzy regions. Assuming that $\Psi \in R^2$ is the vector of non-dimensional performance parameters used to define the operating regions, the identified model for operating region j may be described as a set

$$M_j = \{\Psi_j, V_j, G_j\} \, for \, j = 1, 2, \ldots, N_{op} \qquad (5.1)$$

Where, $\Psi_j = \{\psi_1, \psi_2\}$ is set of non-dimensional parameters used to define an operating region; V_j is a set of fuzzy sets defining an operating region; G_j is sets of identified models for region j. In the generic map shown in Fig. 5.6, two membership functions for parameter ψ_1 and three membership functions for parameter ψ_2 are arbitrarily presumed. The number of fuzzy functions for actual maps relies on the location and number of the operating regions.

Consider a gas turbine with two schedules: region-1 for loads less than 50% of rated value and region-2 for loads in the range of 50–100% of the rated value. In classifying the operating regions, one may select normalized power as ψ_1 and normalized fuel flow rate as ψ_2. Assuming that each operating regions are defined

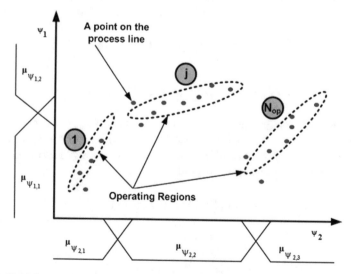

Fig. 5.6 Multiple operating regions for a hypothetical system

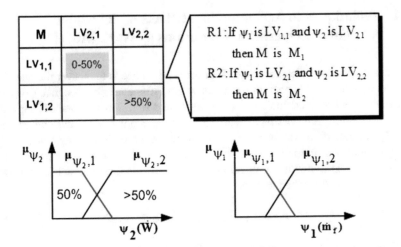

Fig. 5.7 Operating regions for a gas turbine and the corresponding rules

by two Linguistic Variables (LV), Fig. 5.7, the identified models for the two regions
are expressed as

$$M_1 = \{\Psi_1, V_1, G_1\}$$

, and

$$M_2 = \{\Psi_2, V_2, G_2\}$$

Where, $\Psi_1 = \Psi_2 = \{\psi_1, \psi_2\}$; and V_1 are the fuzzy sets involved in each region;
G_1 and G_2 are sets of NF based identified models for the two operating regions,
respectively.

$$V_1 = \{< LV_{1,1}, \mu_{\psi_1,1} >, < L, V_{2,1}, \mu_{\psi_2,1} >\},$$

and

$$V_2 = \{< LV_{1,2}, \mu_{\psi_1,2} >, < L, V_{2,2}, \mu_{\psi_2,2} >\}.$$

The models may include functions for compressor discharge pressure (P_2), power
output $\dot{W}_{ele}$, compressor discharge temperature (T_2), turbine exit temperature (T_7),
models for the gas turbine lubrication system, and models for generator coil temperatures. For the gas path measurements only,

$$G_1 = \{P_{2,1}, \dot{W}_{ele,1}, T_{2,1}, T_{7,1}\},$$

and

$$G_2 = \{P_{2,2}, \dot{W}_{ele,2}, T_{2,2}, T_{7,2}\},$$

To build the FDEM block, several models are needed. Classification of the operating regions in a way described above allows formulating parsimonious models for

any operating region. For instance, the VIGV is not manipulated in the lower load region. Hence, for models that reside to this region, there is no need to include the effect of VIGV. Simani [3] considered the higher load region only while others did not make any distinction between the two regions.

The other factor that needs to be addressed is the effect of external conditions on model accuracy. The load on the gas turbine or cogeneration plant may change over a week due to changing environmental conditions and semester load. As such, the measured parameter trend for one day may not be similar to the trend for next day. In designing a fault detection and diagnosis system, accuracy of the models should not be compromised attributed to neglecting these features. Otherwise, false alarms will be generated while there is none.

The direct way to deal with the changes due to external factors is to collect enough data for each day and concatenate them over the whole year, Fig. 5.8b. To deal with the data size, there are suggested techniques like moving average method and data clustering [4]. These methods reduce the data size before training. Nevertheless, care should be taken to mitigate the drawback of removing important features. In the following, a different approach is proposed.

Consider plant operating data from three consecutive days $\psi_{1,1} \in R^{1 \times N_{d1}}$, $\psi_{1,2} \in R^{1 \times N_{d2}}$, and $\psi_{1,3} \in R^{1 \times N_{d3}}$, selected arbitrarily each data for the same operating region. In the conventional approach, the data sets are combined to form a single vector $\psi_c = [\psi_{1,1}\psi_{1,2}\psi_{1,3}]^T$ and a model is developed based on the new vector. In gas turbines, changes in the external conditions as well as any increase in the load demand are reflected in the power output available at the gas turbine generator terminal. Using the power output as a reference, all the other correlated data can be sorted in ascending order. The effect from data size reduction will be less if the sorted data is used in the process. Figure 5.8 shows plots of hypothetical data set and the difference between the conventional and the newly proposed approach.

For the general case, given $\psi_c = [\psi_{j,1}\psi_{j,2}\psi_{j,3} \ldots \psi_{j,N}]$, the relation between the concatenated ψ_c and the new data set ψ_s is just a mapping

$$\Omega_{\psi_{ref}} : \psi_c \rightarrow \psi_s \qquad (5.2)$$

Where, $\Omega_{\psi_{ref}}$ is a composition function that works by sorting the available data in ascending order based on reference parameter ψ_{ref}.

One advantage of the stated formulation is that, unlike the conventional case the moving average and other clustering algorithms can be effectively exploited to reduce data dimension without compromising main features of the collected data. Gas turbine starting characteristic is mostly having limited number of data points corresponding to each start. Assuming that there are data collected over the week, the proposed approach allows the development of a simple generic model for the start up region. One limitation of the proposed approach is that, it cannot be applied in dealing with vibration data for it is difficult to find a reference parameter.

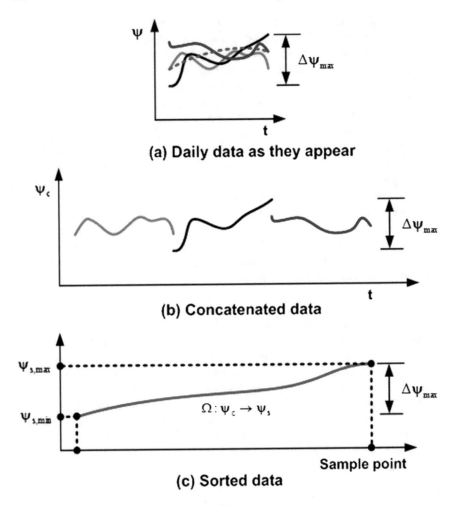

Fig. 5.8 Alternative identification data preparation

5.5 Steps in OBF Based Nonlinear Model Construction

As explained in the discussion of FDEM, the models for the dynamic case and
Normal Operating Conditions (NOC) are developed as NF models that implicitly
use Orthonormal Basis Functions (OBF). One advantage of OBF is that it eliminates autoregressive part. The possible forms of OBFs are elaborated in Chap. 3 and
Sect. 3.2.1. The choice on the specific type of OBF relies on the dynamics of the system. For instance, Laguerre functions are commonly used for a well damped system
with a dominant pole while Kautz functions are suitable for lightly damped systems
with complex conjugate poles [5].

The first step in the formulation of an OBF based model is estimation of the dominant pole and time delay. For a simple system, the pole may be determined from basic laws. For large scale systems, however, adopting such a procedure is quite difficult. In the model suggested by Nelles [5] a two step procedure was applied. In the first step, a model is developed applying ARX model structure. From the step response of the ARX model, the dominant pole and time delay are estimated. In the second step, using the estimated pole and time delay, a nonlinear model in the framework of OBF is developed, Fig. D.3. While Nelles approach is simple to apply, it is only limited to Lagurre model structure. Oliveira et al. [6] and Medeiros et al. [7] argued that in a NF-OBF model the choice of the dominant pole is not crucial if the model order is considered as one of the parameters in the optimization process. In [6], they used pole size of $\xi = 0.7$. Campello and Amaral [8] studied the effect of pole size on MSE while keeping the number of OBFs constant and vice-versa. At the end they chose a value of $\xi = 0.75$. Machado et al. [9] adopted GA to optimize the pole and the whole model structure. In [10], however, a fixed pole size of $\xi = 0.65$ is selected by trial and error. If there is a time delay τ_d known at the start, it can be incorporated in the model by modifying the input as $u(t - \tau_d)$. But, the time delay needs to be estimated separately (see Appendix B.6). In our case, the modelling is done according to the following procedure with the assumption that the pole size and OBFs order are decided by trial and error.

- At first, the input-output data require for both model training and validation are prepared.
- Secondly, suitable model order and pole size are assumed. In the case of GOBF based designs, vector of poles are assumed.
- Using the data from step 1 and step 2, NF-OBF models are trained by the algorithms discussed in Sect. 3.5.
- Finally, the models are validated. The performance criteria outlined in Appendix C.3 can be used to see if the model suffices to provide accurate approximation.

In the OBF based NF model, the Local Linear Time Invariant (LLTI) models in the consequent part of the model rules are all in the form of

$$\begin{cases} \mathbf{g}(p + 1) = \boldsymbol{\Phi}(q, \xi)g_j(p) + \boldsymbol{\Gamma}(\xi)u_j(p) \\ z_k(p) = \sum_{j=0}^{n_u} \theta_j g_j(p) \end{cases} \tag{5.3}$$

For a general nonlinear Multiple Input and Single Output (MISO) model, it can be written as:

$$\begin{cases} \mathbf{g}(p + 1) = \boldsymbol{\Phi}(q, \xi)g(p) + \boldsymbol{\Gamma}(\xi)u(p) \\ y(p) = f(\mathbf{g}(p); \boldsymbol{\theta}) \end{cases} \tag{5.4}$$

In (5.3) and (5.4), $u_j(p)$ stands for the input, θ_j is the parameter to be estimated, $g_j(p)$ is the state, $z_k(p)$ is the output, and ξ is vector of dominant poles of the system.

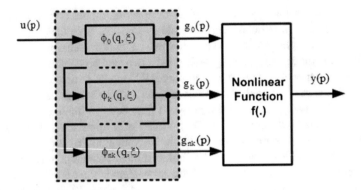

Fig. 5.9 NF model structure in the framework of OBF

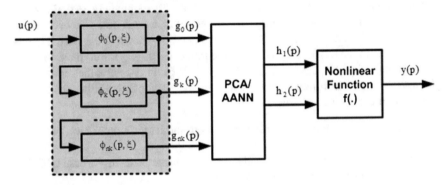

Fig. 5.10 NF model structure in the framework of OBF: with PCA or AANN

The equivalent form of (5.4) is demonstrated in Fig. 5.9. Structure of the matrices $\boldsymbol{\Phi}(q, \xi)$ and $\boldsymbol{\Gamma}(\xi)$ are OBF type dependent, Appendix C.5.

In Fig. 5.10, the nonlinear OBF based model is diagrammed assuming PCA or AANN as input dimension reduction technique.

5.6   New Model Training Algorithm

In this section, we will develop a modified set of equations applicable to recursively train the nonlinear models discussed in Chap. 3. The main motive behind performing this is that, the training algorithms discussed so far have the limitations in real time model updating which is a major setback for online application of nonlinear model based fault detection and diagnosis systems. Replacing the conventional parameter optimization techniques by a recursive approach reduces the training time significantly and hence real-time application of the model to fault detection and diagnosis is possible. In Chap. 3, five types of nonlinear models have been elaborated. In the following, a general form of the models that can be customized to anyone of the five

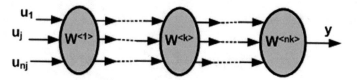

Fig. 5.11 Hypothetical nonlinear model structure

forms is first hypothesized, Fig. 5.11. In the model, matrices corresponding to model parameters in different sections of the nonlinear model topology are represented by $W^{<k>}$ ($k = 1, \ldots, nk$).

The assumed generic equation is

$$y^P = f^P(\mathbf{u}^P; \mathbf{W}^{<1>}, \ldots, \mathbf{W}^{<k>}, \ldots, \mathbf{W}^{<nk>}) + \varepsilon^P \qquad (5.5)$$

Like the previous case, ε^P stands for measurement and modeling errors. The model is assumed time invariant and nonlinear in parameters. What is aimed for is a formulation that is inclined to recursive calculations, which could be using either the Recursive Least Squares (RLS) approach or one of the algorithms featured by bounded error assumption, Chap. 4 and Sect. 4.2. In the following, it is further assumed that matrices of true values of model parameters are designated by $\mathbf{W}^{*<k>}$ ($k = 1, \ldots, nk$) while the approximated parameters are presumed to be arranged in $\hat{\mathbf{W}}^{<k>}$ ($k = 1, \ldots, nk$). What follows is the approximation of (5.5) by Taylor series expansion about the true values. It is necessary that the matrices be reshaped to column vectors. Pertaining to the vector form, a subscript c is added to the matrices name to indicate that it is the vector equivalent. That is, $\mathbf{W}_c^{*<k>}$ ($k = 1, \ldots, nk$) for true values and $\hat{\mathbf{W}}_c^{<k>}$ ($k = 1, \ldots, nk$) for approximations. Neglecting the measurement error and assuming a modeling error that is null, the following can be deduced from (5.5).

$$\hat{y}^P = f^P(\mathbf{u}^P; \hat{\mathbf{W}}^{<1>}, \ldots, \hat{\mathbf{W}}^{<k>}, \ldots, \hat{\mathbf{W}}^{<nk>}) \qquad (5.6)$$

Performing Taylor series expansion on (5.6) leads to (5.7).

$$
\begin{aligned}
\hat{y}^P = f^P(\mathbf{u}^P; \mathbf{W}^*) &+ \left.\frac{\partial(\mathbf{u}; \hat{\mathbf{W}})}{\partial \hat{\mathbf{W}}_c^{<1>}}\right|_{\hat{\mathbf{W}}=\mathbf{W}^*} \left(\hat{\mathbf{W}}_c^{<1>} - \hat{\mathbf{W}}_c^{*<1>}\right) + \ldots \\
&+ \left.\frac{\partial(\mathbf{u}; \hat{\mathbf{W}})}{\partial \hat{\mathbf{W}}_c^{<k>}}\right|_{\hat{\mathbf{W}}=\mathbf{W}^*} \left(\hat{\mathbf{W}}_c^{<k>} - \hat{\mathbf{W}}_c^{*<k>}\right) + \ldots \\
&+ \left.\frac{\partial(\mathbf{u}; \hat{\mathbf{W}})}{\partial \hat{\mathbf{W}}_c^{<nk>}}\right|_{\hat{\mathbf{W}}=\mathbf{W}^*} \left(\hat{\mathbf{W}}_c^{<nk>} - \hat{\mathbf{W}}_c^{*<nk>}\right) \\
&+ O\left(\left\|\hat{\mathbf{W}}_c - \hat{\mathbf{W}}_c^*\right\|\right)^2
\end{aligned}
\qquad (5.7)
$$

Up on defining ξ^p as

$$
\xi^p = f^p(\mathbf{u}^p; \mathbf{W}^*) - \left.\frac{\partial(\mathbf{u}; \hat{\mathbf{W}})}{\partial \hat{\mathbf{W}}_c^{<1>}}\right|_{\hat{\mathbf{W}}=\mathbf{W}^*} \mathbf{W}_c^{*<1>}
$$

$$
- \left.\frac{\partial(\mathbf{u}; \hat{\mathbf{W}})}{\partial \hat{\mathbf{W}}_c^{<k>}}\right|_{\hat{\mathbf{W}}=\mathbf{W}^*} \mathbf{W}_c^{*<k>} + \cdots
$$

$$
- \left.\frac{\partial(\mathbf{u}; \hat{\mathbf{W}})}{\partial \hat{\mathbf{W}}_c^{<nk>}}\right|_{\hat{\mathbf{W}}=\mathbf{W}^*} \mathbf{W}_c^{*<nk>}
$$

$$
+ O\left(\left\|\hat{\mathbf{W}}_c - \hat{\mathbf{W}}_c^*\right\|\right)^2
$$

Equation (5.7) takes the form

$$
\hat{y}^p = \left.\frac{\partial(\mathbf{u}; \hat{\mathbf{W}})}{\partial \hat{\mathbf{W}}_c^{<1>}}\right|_{\hat{\mathbf{W}}=\mathbf{W}^*} \mathbf{W}_c^{*<1>} + \cdots + \left.\frac{\partial(\mathbf{u}; \hat{\mathbf{W}})}{\partial \hat{\mathbf{W}}_c^{<k>}}\right|_{\hat{\mathbf{W}}=\mathbf{W}^*} \mathbf{W}_c^{*<k>} \cdots
$$

$$
+ \left.\frac{\partial(\mathbf{u}; \hat{\mathbf{W}})}{\partial \hat{\mathbf{W}}_c^{<nk>}}\right|_{\hat{\mathbf{W}}=\mathbf{W}^*} \mathbf{W}_c^{*<nk>} + \xi^p \tag{5.8}
$$

If one assumes that true values of the model parameters are known and ξ^p is neglected, (5.8) becomes a linear in parameter model. This is very helpful in recursive calculations. By presuming that the parameters at the preceding estimation are known, a new set of parameters can easily be estimated. In the following, instead of considering (5.6), we will consider (5.8) to characterize the actual system. If (5.8) is substituted in (5.5), the resulting equation leads to

$$
y^p = \boldsymbol{\Psi}^p \boldsymbol{\theta} + \gamma^p \tag{5.9}
$$

Where, $\boldsymbol{\Psi}^p = \begin{bmatrix} \dfrac{\partial(\mathbf{u}; \hat{\mathbf{W}})}{\partial \hat{\mathbf{W}}_c^{<1>}} & \dfrac{\partial(\mathbf{u}; \hat{\mathbf{W}})}{\partial \hat{\mathbf{W}}_c^{<k>}} & \dfrac{\partial(\mathbf{u}; \hat{\mathbf{W}})}{\partial \hat{\mathbf{W}}_c^{<nk>}} \end{bmatrix}$; and

$$
\boldsymbol{\theta} = \begin{bmatrix} \hat{\mathbf{W}}_c^{<1>} & \hat{\mathbf{W}}_c^{<k>} & \hat{\mathbf{W}}_c^{<nk>} \end{bmatrix}^T
$$

The error term γ^p is the sum of the errors due to model approximation and error resulting from input-output measurements. The matrices associated with the five models are mentioned in Table 5.1. Refer to Appendix B for the derivatives of the model output with respect to the model parameters and corresponding to each of the models. Note that AANN is not included in the list for it is a MIMO model and finding derivatives of the model outputs with respect to the model parameters is quit time consuming but not difficult.

Equation (5.9) is in a suitable form that can be solved by well know recursive algorithms. In the conventional approaches, the sum of squares of the errors between

Table 5.1 Parameter matrices involved in computational intelligence based models

Model type	Involved matrices
Feed Forward Neural Network (FFNN)	$\tilde{\mathbf{W}}^{<1>}, \tilde{\theta}$
Radial Basis Function (RBF) Neural Network	$\mathbf{C}, \Sigma, \tilde{\theta}$
Normalized Radial Basis Function (NRBF) Neural Network	$\mathbf{C}, \Sigma, \theta$
Neuro-Fuzzy Singleton Model (NF-S)	$\mathbf{C}, \Sigma, \theta$
Neuro-Fuzzy (NF-TSK) Model	$\mathbf{C}, \Sigma, \mathbf{W}^{<5>}$

the model and actual data are minimized to estimate the model parameters with the prior assumption that measurement and model errors are iid. An alternative and robust approach is to assume that the errors between the parametric model and the actual data are bounded in a certain region. The latter technique is known as parameter set or feasible set estimation technique. The parameters are estimated in such a way that they are consistent with the measurements and error bounds. But, actual shape of the resulting feasible set is a polyhedra and finding its final shape and size applying linear programming procedures is a cumbersome procedure. Often, instead of directly dealing with the polyhedra, simple structures like Ellipsoids and orthotopes are used to approximate the feasible set. For solving (5.9), the Ellipsoid algorithm as discussed in Sect. 4.2 is applied. Besides, rather than initializing the network parameters assuming all zero, random initialization applying *rand* function from Matlab is considered. The use of the proposed approach and its performance as compared to the other techniques will be presented in Chap. 6.

5.7 Summary

In this chapter, the last part of the research methodology was considered. It described the design of an improved fault detection and diagnosis system using the concepts elaborated in Chaps. 3 and 4. A summary of the main contributions from this chapter are given as follows:

- In Sect. 5.2 of the chapter was presented the proposed IFDD system, which is the first contribution. The design is featured by three sequentially connected modules: DPM, FDEM, and FDIM. In the structure of the DPM, DWT and PCA are assumed to deal with measurement noise and data dimension reduction, respectively. An approach relying on NLPCA is also included in case if PCA fails to provide the expected performance or if there be a need to make comparision test in the next chapters. In the FDEM, the design catagorizes the module into four groups input sensors, output sensors, soft sensors, and sensors for nonparameteric models which allows detail characterisation and the use of all available signals. It is not only the structure that is unique; it supposed OBFs in the NF structure for the case where the

FDEM is needed to function in dynamic conditions. Besides, the detection system is designed in such a way that adaptive model uncertainity equations are the heart of the whole procedure. The FDIM triggered by the result from FDEM acts upon the residuals to provide conclusions. The analysis procedure in the FDIM is structured with two possible options: binary and fuzzy. In Sect. 5.3, the subsystem level FDD is generalized to overall plant level FDD system.

- The second main contribution from the current chapter resides in Sect. 5.4. A fuzzy approach is proposed to deal with multiple operating regions of a system. The idea of treating multiple operating regions is introduced for it permits realistic handling of the situation.
- In Sect. 5.5 was discussed the third contribution that refers to the design of NF models in the framework of OBFs. In fact, instead of just focusing on NF, the discussion was made in a sort of general way that allows the use of other nonlinear models as well. The last part of the discussion focussed on the estimation of time delay, time constant and dominant pole from a step response of the system.
- The last contribution was documented in Sect. 5.5 of the chapter. It was proposed a modified and fast model training algorithm that makes use of the recursive techniques elaborated in Chap. 4.

Finally, it can be concluded that the present chapter well addressed the design of an improved FDD system applying the concepts given in Chaps. 3 and 4. Basically, it fulfilled the main objective of the research. In the next chapter, validation of the proposed ideas will be presented.

References

1. Kramer, M. A. (1991). Nonlinear principal componenet analysis using autoassociative neural networks. *AIChE Journal*, *37*, 233–243.
2. Scholz M., Fraunholz M., Selbig J. (2007) Nonlinear Principal Component Analysis: Neural Network Models and Applications in Principal Manifolds for Data Visualization and Dimension Reduction, pp. 44–67.
3. Simani, S. (2005). Identification and fault diagnosis of a simulated model of an industrial gas turbine. *IEEE Transactions on Industrial Informatics*, *1*, 202–216.
4. Guo Z., Uhrig R. E. (1992). Using genetic algorithms to select inputs for neural networks. In *COGANN-92: International Workshop on Combinations of Genetic Algorithms and Neural Networks*. pp. 223–234.
5. Nelles, O. (2001). *Nonlinear System Identification*. Berlin: Springer.
6. Oliveira, G. H. C., Campello, R. J. G. B., & Amaral, W. C. (1999). Fuzzy models within orthonormal basis function framework. In *Fuzzy Systems Conference Proceedings, FUZZ-IEEE '99. IEEE International*, *2*, 957–962.
7. Medeiros, A. V., Amaral W. C., Campello R. J. G. B. (2006). GA Optimization of OBF TS Fuzzy Models with Linear and non Linear Local Models. In *2006 Ninth Brazilian Symposium on Neural Networks (SBRN '06)*. (pp. 66–71).
8. Campello R. J. G. B., Amaral W. C. (2002) Takagi-Sugeno fuzzy models within orthonormal basis function framework and their application to process control. In *Proceedings of the 2002 IEEE International Conference on Fuzzy Systems, FUZZ-IEEE'02*. (pp. 1399–1404).

9. Machado J. B., Amaral W. C., Campello R. J. G. (2007). Design of OBF-TS Fuzzy models based on multiple clustering validity criteria. In *19th IEEE International Conference on Tools with Artificial Intelligence (ICTAI 2007)*. (pp. 336–339).
10. Alci, M., & Asyali, M. H. (2009). Nonlinear system identification via Laguerre network based fuzzy systems. *Fuzzy Sets and Systems.*, *160*, 3518–3529.
11. Saravanamutto H. I. H., Rogers G. F. C., Cohen H. (1996). *Gas Turbine Theory*, (4th ed.).: Longman Group Limited.
12. Walsh P. P., Fletcher P. (2004) *Gas Turbine Performance*. Blackwell Science Ltd.
13. Schobeiri M. (2005) *Turbomachinery Flow Physics and Dynamic Performance*. Verlag, Berlin, Heidelberg: Springer.
14. Razak A. M. Y. (2007). *Industrial Gas Turbines Performance and Operability*. Woodhead Publishing Limited, CRC Press, LLC.
15. Chui C. K. (1992). *An Introduction to Wavelets*. Academic Press Limited.
16. Nelles, O. (1997). Orthonormal Basis Functions for Nonlinear System Identification with Local Linear Model Trees (LOLIMOT). *IFAC Symposium on System Identification (SYSID)* (pp. 667–672). Fukuoka, Japan: Kitakyushu.
17. Haglind, F. (2010). Variable geometry gas turbines for improving the part-load performance of marine combined cycles - gas turbine performance. *Energy*, *31*, 467–476.
18. Coverse G. L. (1984). *Extended Parametric Representation of Compressor Fans and Turbines* Volume 2: Part user's Manual (Parametric Turbine). NASA-CR-174646.
19. Seborg D. E., Edgar T. F., Mellichamp D. A. (2003). *Process Dynamics and Control*. Wiley, New York.
20. Mikles J., Fikar M. (2007). *Process Modelling Identification and Control*. Verlag: Springer.
21. Srinivasarao M., Patwardhan S. C., Gudi R. D. (2006). From data to nonlinear predictive control. 1. Identification of multivariable nonlinear state observers. *Industrial and Engineering Chemistry Research*.
22. Dincer I., Rosen M. A. (2007). *Exergy: Energy, Environment, and Sustainable Development*. Elsevier Ltd.

Chapter 6
Application Studies, Part-I: Model Identification and Validation

6.1 Introduction

In Chaps. 4 and 5, all the theories needed for developing fault detection relevant models in either static or dynamic conditions, and a method for estimating their uncertainties, respectively, have been thoroughly discussed. The good side about the presented approaches is that they are all nonlinear in parameter models. Even more interesting is the dynamic models in the framework of OBFs. In this chapter, the application of the proposed theories to a specific Cogeneration and Cooling Plant (CCP) is presented. First, detail description of the CCP is provided in Sect. 6.2. Secondly, the use of the static and dynamic models developed for the critical sub-systems GTG, HRSG and SAC are investigated in Sect. 6.3. In the same section are contained analyses of the effect of model structure, error assumption and optimization algorithms on model uncertainty. Validation of the semi-empirical models for the GTG and HRSG are also included. Finally, Sect. 6.4 provides a summary of the contribution from this chapter.

6.2 Description of the Sub-Systems in the CCP Selected for the Case Study

The CCP considered for the case study is from Universiti Teknologi PETRO- NAS. The whole plant is equipped with two CCPs with a common Steam Header (SH), Fig. 6.1. The electric power from the two systems is connected to a system bus from which load shading is performed. It can also be recognized from Fig. 6.1 that the steam from the two cogeneration systems (COGEN-I and COGEN-II) is discharged to a steam header from which controlled flow is supplied to the SACs. When the cooling load demand is high and the electric load is low, the steam from one COGEN drives both SACs. The cooling water required for operating the SACs is supplied

© Springer International Publishing AG 2018
T.A. Lemma, *A Hybrid Approach for Power Plant Fault Diagnostics*,
Studies in Computational Intelligence 743,
https://doi.org/10.1007/978-3-319-71871-2_6

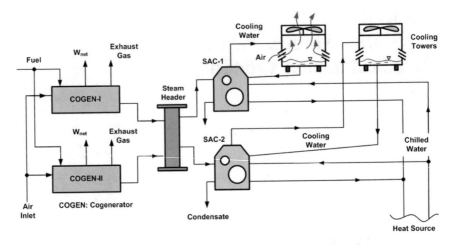

Fig. 6.1 Schematic diagram of the cogeneration and cooling plant considered for case study

from cooling towers at constant flow rates and constant supply temperature. In the sequel, half of the whole setup that includes one COGEN and an SAC is considered for case study.

6.2.1 Gas Turbine Generator

The gas turbine generator, Fig. 6.2a, which is part of the CCP considered for the case study, is Siemens designed Solar Taurus 60 s engine. The thermodynamic cycle governing its operation is as shown in Fig. 6.2b. The atmospheric air is first compressed in the multi stage axial compressor (process 1–2). The compressed air then flows through the diffusion duct and latter fuel added and burnt in the combustion chamber (process 2–3). The hot gas leaving the combustion chamber expands through the turbine (process 3–4) resulting in power enough to drive the compressor, auxiliary rotating parts connected to the turbine shaft and the main generator rotor. In addition to the basic components, the gas turbine is equipped with VIGV, VSVs, SIVs, and a bleed duct with a Bleed Valve (BV). In service, it is featured by three modes of operation start up, operation with a load less than 50% and load greater than 50% of rated capacity, respectively. Caused by malfunctions, wrong calibration of the measuring instruments or ageing, a fault or failure may occur in one of the three operating regions.

The starting mode of operation is featured by a sequence of operations. At first, the start system gradually drives the engine from standstill to the purge speed, which is around 20% of the synchronous speed. Once it reaches to the 20% speed, it stays there for about 20–25 s, Fig. 6.3. The reason is to purge the engine with a clean air that will avoid explosion during engine ignition. After the purging process is

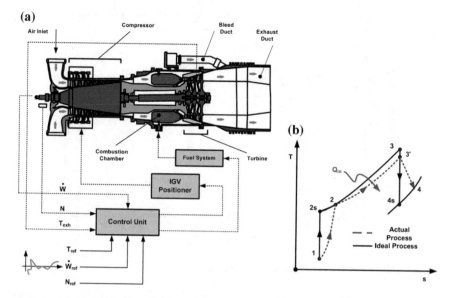

Fig. 6.2 Single shaft gas turbine generator: **a** Schematic Diagram, and **b** Process diagram

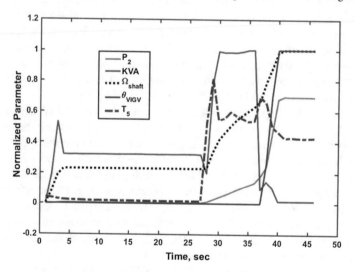

Fig. 6.3 GTG start cycle as a function of time

completed, the engine will be lit-off and accelerated to the synchronous speed. Until the engine reaches a self-sustaining speed, which is around 65% of the synchronous speed, part of the accelerating torque is provided by the starting system, Sect. 6.2.1. At 65% speed, the start system will disengage and the turbine will run by itself to the synchronous speed.

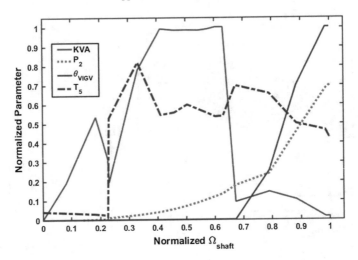

Fig. 6.4 GTG start cycle as a function of shaft speed

It is not only the start system that is involved in the process. In the gas turbine generator, compressor surge is a phenomenon that needs to be avoided. If is happens, the engine will fail to reach to the full speed. There may also be failure of the compressor blades as a result of pulsating nature of the pressure inside the compressor. To combat this unwanted behaviour, the VIGV, VSVs and BV are manipulated. For speeds between 0 and 65%, the VIGV is closed and the VSVs are at the design stagger angles. When the speed climbs from 65% towards 100%, the VIGV and VSVs are opened to gradually increase the air flow, Fig. 6.4.

In the first mode of operation, where the load is less than 50% of rated load, the VIGV is fully open and the GTG is on load control. That is to say the controller only varies the fuel flow rate to meet the power demand with the shaft speed remain constant. In the second mode of operation also called SoLoNOx mode load greater than 50% of the rated load the VIGV and the stators in the first three stages of the compressor are manipulated to maintain set point temperature of the exhaust gas at the inlet to the third stage of the gas turbine. At combustion temperatures higher than 1593 °C the emission of NOx is higher. At combustion temperatures lower than 1482 °C, the emission of CO is higher. The SoLoNOx is a lean premix combustion system designed to avoid the high emission regions.

The SoLoNOx mode of operation starts at 50% load. At this point the VIGV is closed as necessary to meet the set point temperature T_5. In the GTG considered in this case study, the set point temperature is about 667 °C. As the load increases, the VIGV is allowed to open as required to keep T_5 set point temperature. When the load on the GTG is higher to push the temperature T_5 above the set pint, the VIGV is fully open and the Swirl Inlet Valves (SIVs) are closed to reduce emission levels. As the load increases further, the SIVs are opened in a controlled manner to meet emission limits. In the GTG under consideration, the third region where SIVs are manipulated is not common for the load requirement is mostly less than 4.2 MW while the design capacity is 5.2 MW.

The GTG is also featured by compressed air connections. One such piping is the line connecting the aft end of the compressor with the bearing lube oil chambers. In the GTG, air-pressurized labyrinth seals are used to prevent leakage of lube oil from the bearings. And, the required air quantity is tapped from the compressor aft end. For cooling the duct assembly, combustor and rotor blades, compressed air from compressor diffuser is extracted. The first stage turbine rotor disc and first stage rotor blades are cooled by the compressed air tapped at the exit of the compressor diffuser. Compressor bearing support housing, gas producer centre bolt and rotor disc are cooled by the air leaked from the 11th stage of the compressor. Part of the cooling air extracted at the exit of the compressor diffuser is also directed to the first stage turbine nozzle. The air leaves the nozzles at the trailing edge.

The gas turbine is coupled to a generator through a planetary gearbox and it runs at constant speed governed by the line frequency. Even though the ambient conditions and the load are changing, the speed remains the same and equals to the synchronous speed. For 60 Hz line frequency, the speed at the generator shaft is 1800 RPM while it is 1500 RPM for 50 Hz operating frequency.

Droop Verses Isochronous Modes of GTG Operation

As elaborated in Sect. 6.2, the electric power output from each COGEN is passed to the system bus before dispatching is done. In the case of either high cooling load demand or electric power demand or both, the two COGENs run together. As such, in covering the electric power load, one gas turbine operates in cooperation with another. This is possible because of the setting options at the control panel that allows two modes of operation for a GTG. Depending up on the setting from the control panel, one may work in droop mode while the other is running in isochronous mode. It is important to consider the distinction between these modes of operations for they affect reliability of the identified models in designing fault detection and diagnosis strategies.

If the controller is set to isochronous mode, the controller keeps the shaft speed constant regardless of the load change. This mode of operation is very important for there is restriction in the line frequency variations, especially if the gas turbine is connected to a grid. The second mode of operation, droop mode, maintains the shaft speed as a function of the load. A maximum of 5% is the common setting for allowed shaft speed variation. When the setting in the turbines is to run one gas turbine in isochronous mode and the other in the droop mode, any additional load coming to the system is first taken care by the isochronous turbine. Latter the added load is transferred to the droop turbine. To avoid power trip, however, the difference between the powers produced by the turbines is limited to 500 kW. In this case any additional load requirement in the range of 0–500 kW is taken care of by the isochronous turbine. In cases where the demand increase is higher than 500 kW, the load setting for the droop turbine will be readjusted. A plot of actual electric power at the generator terminals for two gas turbines working cooperatively is indicated in Fig. 6.5. As the figure shows more variation of load is experienced by the isochronous turbine. Latter in the model identification, data from the isochronous operation will be considered for further analysis.

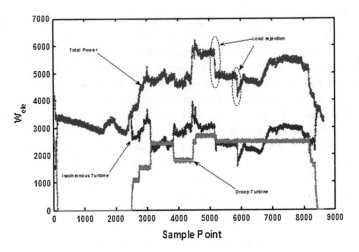

Fig. 6.5 Actual operation trend for single-shaft gas turbine generators

GTG Auxiliary Systems

The complete GTG is also featured by auxiliary systems. Some of them include: air intake system, exhaust gas system, enclosures, electrical and control system, lube system, and start system. The lube system defined by separate piping, gas turbine driven primary oil pump, AC motor driven pre/post oil pump, oil filters, coolers, measurement gauges, regulators and strainers provides lube oil to the gas turbine and generator. The heat and any debris in the power transmission system are removed by the oil. The bleed valve, VIGV and VSVs actuator, and fuel actuators are driven by a hydraulic force. The source of this force is the oil from the same system. The gas turbine shafts are supported by hydrodynamic bearings where an oil film separates the rotating shaft and the stationary support. The required oil for the hydrodynamic bearing is furnished by the same lube oil system. In consideration of the stated points, it can be said that the lube oil system is a very critical part of the system. Reduced oil pressure due to leakage or failed pressure regulators may lead to contact between rotating and stationary parts, the end result being costly bearing damage and long down-time. The performance of the oil is also a function of temperature, which affects viscosity of the oil. Failure to keep the required temperature caused by faults in the cooling system has the consequence of deteriorating the oil life.

In the GTG, the temperature and pressure of the lube oil is monitored at selected locations. The schematic diagram demonstrating the bearing locations and the lubricant path is given in Fig. 6.6. Similar to the gas path systems, it also needs models for normal operating conditions. Abnormal conditions could be detected making use of the NOC models. Unique feature of this system is that measured variables are difficult to classify as input and output. Besides, the operating trend varies from one day to the other due to the load variation. This necessitates a data based model covering the whole operating region.

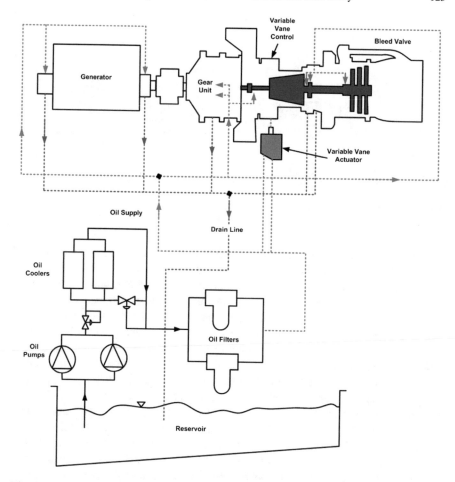

Fig. 6.6 Gas turbine generator lubrication oil path and bearing locations

The other auxiliary system worth discussing is the start system. There are many ways to start a gas turbine. Pneumatic start, hydraulic start and starting by an AC motor are the common methods. In the case that is considered under this chapter, the GTG is designed with a Variable Frequency Drive (VFD) AC motor. The motor and the gas turbine shaft are connected through a sprag clutch assembly. Schematic diagram demonstrating basic parts of the start system is indicated in Fig. 6.7. The starting sequence in a gas turbine is a complex process. It involves ramping to 20% speed, purging or cranking for a preset number of seconds at 20% speed, acceleration by starter motor up to self-sustaining speed (65% speed) and ignition accompanied by VIGV and VSVs opening until the shaft speed reaches to the synchronous speed. During starting, the variables considered measurable in the loaded region are also measurable during GTG starting. The additional parameters

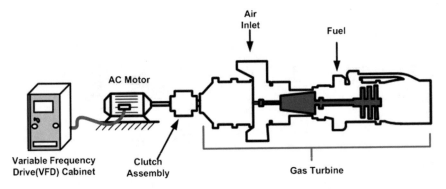

Fig. 6.7 GTG start-up system

available for measurement and corresponding to this operating region are speed of the starter shaft, current to the starter motor and VFD generated voltage.

The faults often encountered during start-up are flame out, hung start and hot start. Flame out is the extinction of the flame in the combustion chamber caused by interrupted fuel flow. Failure in the control system may lead to this fault. Hung start is the starting process being slowed that may be caused by starter motor malfunction. If the toque developed by the starter motor is not enough, the gas turbine will experience a slow acceleration which is to be avoided for vibration related reasons. On the other hand, hot start is evidenced by high gas temperature at the inlet to the gas turbine.

Prediction models are developed for starter motor shaft speed, turbine shaft speed and temperature of the hot gas at the inlet of the 3rd stage of the gas turbine. The input presumed for developing the models are starter current, starter voltage, fuel flow, and VIGV position. For each case six days data are used for model training while two days data collected after one week latter are used for model validation. The decision on the number of days relates to the need to avoid the effect of GTG performance drop caused by compressor fouling.

Challanges in Developing a High Fidelity Mechanistic Model for the Gas Turbine Generator?

The procedures for simulating a gas turbine generator at the design and off-design conditions are well documented in the books by Saravanamutto [1], Walsh and Fletcher [2], Schobeiri [3], and Razak [4]. All the methods, however, rely on the design information that is hardly available for proprietary reasons. The following are true about the GTG under consideration.

- The GTG works under the action of a controller with a built in low value selector whose inputs are three control signals: temperature T_5, shaft speed Ω_{sh}, and electric power output $\dot{W}_{ele}$. None of the controller and measurement sensor gains and time constants is available.
- The axial compressor is embedded with VIGV and VSVs linked to a driving mechanism, Fig. 6.8. The VIGV and VSVs are adjusted such that the relative flow

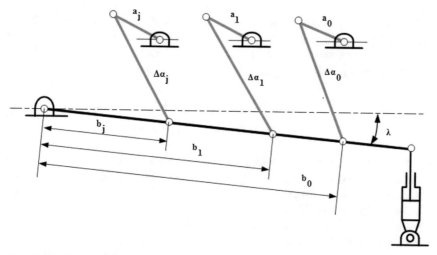

Fig. 6.8 VIGV and VSVs driver link mechanism

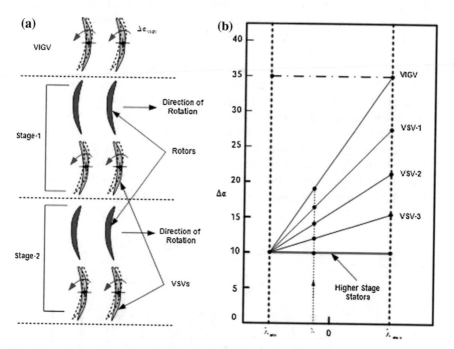

Fig. 6.9 Variation of VIGV and VSVs angle setting

angles are constant and equal to the stagger angles at the design point. Hence, the performance of a stage varies with the stagger angle of VIGV and each VSV, Fig. 6.9. Unless the dimensions of the driving links and the blade geometries are known, it is a far-fetched situation if one thinks that the performance of the variable compressor is accurately modeled.

- The reaction inside the combustion chamber is governed by Eq. (6.1). The fuel is a natural gas but the values of x, y and z are not known exactly. These values are critical in the estimation of adiabatic flame temperature and specific heat capacity of the exhaust gas. λ is the excess air coefficient.

$$C_xH_yO_z + \lambda \left(x + \frac{y}{2} - z\right)(O_2 + 3.773N_2)$$
$$\rightarrow aCO_2 + bH_2O + cO_2 + dCO + eN_2 + \cdots$$

(6.1)

- One of the auxiliary systems in a gas turbine is the air system. The purpose of the air system is to provide cooling air to the hot sections, bearing chamber, and controller circuits of the gas turbine. The air required for cooling and other purposes are extracted at a certain location in the compressor with the condition that the pressure is sufficient enough to overcome the losses in the flow path and the static pressure at the sink point is higher than the surrounding static pressure. The amount of air extracted from the compressor affects the performance of the system. Referring to the data given in Walsh and Fletcher [2], for turbine disc cooling and rim sealing a quantity of 0.5% per disc face is recommended. For bearing chamber sealing, around 0.02 kg/s per chamber is recommended. In addition to the two flows, a separate flow is required for cooling first stage stators. The amount for cooling the first stage stator vanes and rotor blades is often considered as technology dependent. The turbine in the GTG is a cooled turbine with the cooling air tapped form 11th stage of the compressor and at the outlet from the diffuser. The assumption of a single performance map for the whole turbine overlooks the effect of cooling air circulation inside the rotating blades and the stationary nozzles. Hence, it is easy to recognize that, the detail geometry of the turbine components and governing design values for the cooling air system need to be known. For the GTG under investigation, there is no data available except the description on the path of flow. Figure 6.10 shows the cooling air and main flow network.

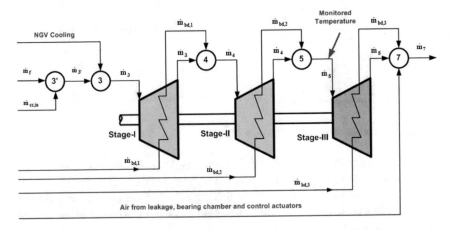

Fig. 6.10 Cooling air path in the gas turbine generator

- The GTG is also feature by ducts at different locations. In fact, the combustion chambers may also be considered as ducts with non adiabatic condition. Depending up on the location of the duct, there is a possibility of total pressure loss. To take this effect into account while performing off-design calculation, the loss magnitude at the design point need to be known.

From the above mentioned points, it can be said that one may develop a steady state or transient model for a GTG relying on rough assumptions. However, since our objective here is to develop a model accurate enough to detect changes due to error in the measurements as well as performance derangements caused by process faults, accuracy of the model is critical. In fact, for the mechanistic model the computational time remains high that will make it impractical for the use of the model in parallel with the fault detection procedure.

6.2.2   Heat Recovery Steam Generator

The second unit that defines the CCP is a Heat Recovery Steam Generator (HRSG). The main reason for having the HRSG is to recover the heat from the exhaust gas that otherwise will be discharged to the environment. The addition of this unit results in an increase in the thermal efficiency of the system. In Fig. 6.11a, we find schematic representation of the HRSG. As indicated, the HRSG is a natural circulation drum boiler. The quantity of exhaust gas flowing through the HRSG is controlled by a diverter damper. The exhaust gas, when passing through the HRSG, heats the riser tubes, down-comers and economizer tubes. Since the feed water is flowing inside the tubes, the heat transfer by convection and conduction will raise the temperature of the feed water to a level that will cause the water to vaporize. In the steam drum the steam is separated from the hot water. In case the concentration of foreign matters in the steam drum goes beyond the set limit, controlled quantity of hot water from the steam drum is discharged to the drain system through the blow-down heat exchanger. The temperature profiles for the exhaust gas and feed water, respectively, are shown in Fig. 6.11b. Known design point parameters of the HRSG are: - steam flow rate ($\dot{m}_s = 12$ Ton/h), exhaust gas inlet temperature ($T_{g1} \sim 540\,^{\circ}\mathrm{C}$), exhaust gas flow rate ($\dot{m}_{exh} = 71826$ kg/h), and working pressure ($P_{wp} \sim 1176.8$ Pa).

In the design of the HRSG, provisions are available to keep the steam drum water level in a certain limit specified by the manufacturer. If the water level is above the maximum set point, then there will be water carry over that will distract safe operation of the system on the discharge side. On the other hand, water level less than that permitted for safe operation will expose the tubes that may lead to rapture and latter failure of the system. The two measures often available to get around these phenomena are the steam pressure controller and the drum water level controller, Fig. 6.11a. The pressure controller takes signal from the Pressure Transmitter (PT) and manipulates the diverter damper position to keep the set point pressure. Regarding the water level controller, the HRSG is designed to operate in two control modes

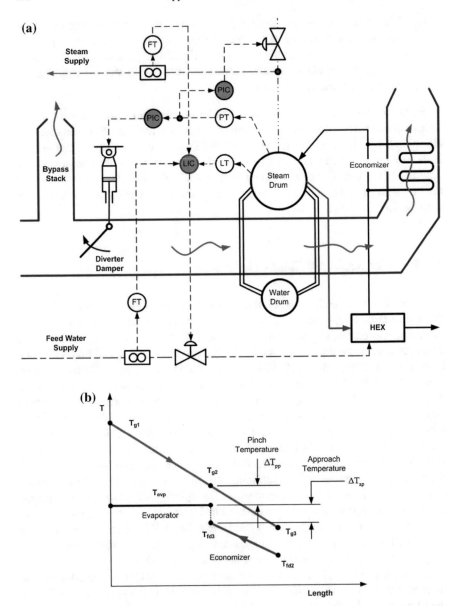

Fig. 6.11 Schematic and process diagrams for the HRSG: **a** Simplified schematic diagram, and **b** Approximate temperature profile

depending up on the system duty. In the first mode that refers to HRSG load less than 30% of the design duty, the HRSG is run by a single element controller. As such, the feedback signal to the controller is only from the Level Transmitter (LT). For higher loads, however, the HRSG runs under three-element controller. During this time, the controller decides the opening of the feed water valve position based on input signals from the water LT, steam Flow Transmitter (FT), and feed water FT. The main reason behind using three-element controller is to avoid the false water level signal that may be caused by swelling or shrinkage of the water level in the steam drum.

Swelling and shrinkage of the steam drum water level can be explained as follows. When more steam is discharged from the steam drum, the pressure in the steam drum decreases. The reduced pressure then leads to more bubbles formed inside the drum. This creates a tendency of the water level to increase. This phenomenon is called swelling. A single element controller having this signal will be deceived and try to reduce the feed water quantity. In the other scenario, when the steam discharge is reduced, the pressure inside the drum increases and hence the water level gets lower resulting in the other effect termed as shrinkage. A single element controller having this signal tends to increase the feed water flow while the actual requirement is to decrease the flow rate. As mentioned earlier, these are in fact the reasons behind having a three-element controller. The three-element controller gets feedback signal from three sources: feed water FT, steam FT and drum water LT, Fig. 6.11a. That way it compensates for the wrong signal.

It can be said that, actual development of a mechanistic model for the HRSG is complicated by the difficulty to know the values of the controller gains, valve dynamics, pinch temperature, approach temperature, steam drum size, riser and down-comer pipes diameter, and details if there is pressure compensation or specific configuration of the controller. As a solution to the need to a reliable model for FDD system design, we intend to rely on data driven models.

6.2.3 LiBr-H2O Steam Absorption Chiller

The third sub-system in the CCP is the Steam Absorption Chiller (SAC). The SAC is used to cool the water sought for air-conditioning. The SAC is chosen for it can be driven by a hot medium at temperatures in the range of 100 to 200 °C. There are different designs of SAC but the chiller used in the CCP is a series, double effect, LiBr/H_2O design in which water is used as a refrigerant. The chiller runs relying on three flow streams. The first stream is the chilled water supply and return system. Detail schematic diagram of the SAC and the corresponding design temperatures are available in Appendix F. The corresponding simplified schematic and diagram are given in Fig. 6.12. The hot water coming from the buildings passes through the evaporator where it looses the heat to the refrigerant (H_2O). The driving energy to effect heat removal from the cold space (evaporator) is obtained from the steam supplied at the high pressure generator. The water vapor or refrigerant from the evaporator flows

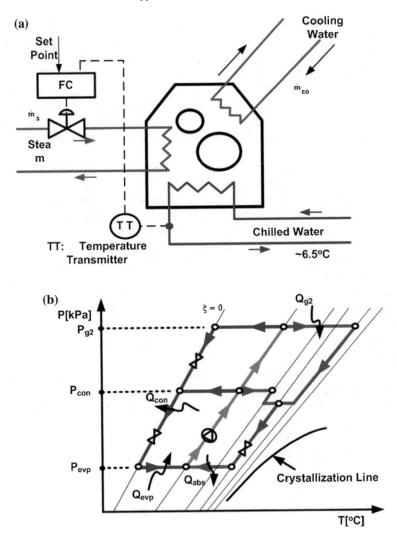

Fig. 6.12 Simplified schematic and diagram for the SAC: **a** Simplified diagram, and **b** Duhring diagram for the SAC

to the absorber. In the absorber, the refrigerant and the strong solution of LiBr-H_2O mix exothermically. The resulting weak solution is pumped to the low pressure and high pressure generators with the flow split ratio set according to the design and control actions. For efficiency related reasons, the weak solution passes through the high temperature, low temperature and drain heat exchangers, respectively, before it reaches to the generators.

Temperature of the supplied steam at the high pressure generator is above $160\,^{\circ}\text{C}$. Because of the high temperature, the energy content of the vapor generated in the high pressure generator is enough to run the low pressure generator. After the LiBr-H_2O solution completes the process in the two generators, it flows back to the absorber through the heat exchanger, mixer and expansion valves. To complete the chiller cycle, the heat generated in the abosber and the heat that needs to be expelled from the system at the condenser are removed by the cooling water supplied from the cooling towers (not indicated in the figures). The cooling water system is the third stream required to define the chiller system. The electricity requirement by the chiller is a minimum for the only power required is to drive the solution pumps and corresponding control systems.

Considering control of the chiller, it is designed to operate for loads in the range of 20–100% of design duty. Regardless of the load, the chilled water supply temperature is controlled to about 6.5 °C. The cooling water flow rate and supply temperature do not vary much. The parameter that is varied to meet the chilled water supply temperature regardless of the return temperature is the flow rate of steam supply. However, because of the concern on crystallization, when the system is operating at low loads, the solution flow rate is also manipulated automatically.

Similar to the case of GTG and HRSG, detail characterization of the SAC by thermodynamic equations is difficult because of many of the design parameters not available for users. The SAC is not a new one. It has been in use since 2004 with servicing and overhaul performed every monthly and annually. During this period, the heat transfer characteristics of the component are not like the values at the design stage. For complete modeling, dynamic characteristics of flow control valves and pump characteristics need to be available. In fault detection and diagnosis, modeling error will be wrongly classified as real fault if the models for normal operating conditions are not appropriate. The solution proposed in this thesis is to rely on measured data to develop updatable models. To get the feel of relative operating points, the partially available design point information, Appendix E, will be used for the purpose of normalizing the measured data prior to model training and validation.

6.3   Validation and Analysis of Nonlinear Models for Normal Operating Conditions

This section presents the identified models from actual data of a cogeneration and district cooling plant. In the first part of the presentation, the models for GTG are given. Following that, similar studies are made on the HRSG and SAC. Instead of just developing the models, however, the approach is first creating models and then analyzing them in terms of how the models respond to different optimization decisions. We try to expose these through model uncertainty and computation time. It is worth stressing that the whole attempt would be done with due attention that the final models would be aimed at fault detection and diagnosis.

Considering the gas turbine as elaborated in Sect. 6.2.1, which is the first part of the cogeneration plant, several measurements are acquired from the SCADA system. For a gas turbine load up to ~50% of the rated value, the VIGV position is held fully open. For this reason, the input considered for the gas path system is only the fuel flow rate. For loads higher than 50% of rated quantity, the VIGV, VSVs and the fuel flow rate are manipulated to meet the load and high temperature exhaust gas requirements. Because the models are for normal operating conditions, the measurement errors $(\tilde{u}(p), \tilde{y}(p))$ and sensor faults $(f_u(p), f_y(p))$ are set equal to zero. In the following section, static and dynamic models will be developed for the three critical subsystems.

6.3.1 Models for the Gas Turbine Generator

The CCP from Universiti Teknologi PETRONAS is equipped with a power management system. For the GTG alone, a dedicated Turbotronic Control system is available. To this system all the measurement sensors are connected through a serial cable. With the help of the system monitoring computer and the Windows based Solar GTG software, one can access historical data on the operation of the gas turbine. The dedicated computer allows USB connection and that made it possible to copy the daily data and use it in the fault detection and diagnosis algorithm. More than 82 readings are available for trending, each collected in 10 s interval. The measurements selected for the case study are partly shown in Chap. 2 and Fig. 2.2. The rest of the signals included in the models are listed in Tables F.5 and F.6 of Appendix F. The signals from the data management system are first de-noised applying suitable levels of DWT. For instance, Fig. 6.13 demonstrates the raw data and the pre-processed data corresponding to the temperature at the inlet of the third-stage of the gas turbine, T_5. A DWT of Daubechies [5] type with level-5 is used to reduce the noise. A different level is also possible with the limitation that higher levels may eliminate important features of the data.

The next major step in NF or ANN based model identification is data normalization. In this book the GTG data are normalized about partially known design point values and maximum alarm setting.

GTG Models for Operating Loads Lower than 50% of Rated

Adopting the concepts proposed in Sect. 5.4, the less than 50% load region (calling it operating region-1) is defined by fuzzy set V_1 (refer to Fig. 5.7). Based on (5.1), the model set that defines the operating region is

$$M_1 = \{\Psi_1, V_1, G_1\}$$

where, $G_1 = \{G_{1,1}, G_{1,2}, G_{1,3}, G_{1,4}\}$; $G_{1,1}, G_{1,2}, G_{1,3}, and \ G_{1,4}$ are model sets for the sensors in the gas path, lubrication system, generator coils and temperature sensors at the inlet to the 3rd stage of the gas turbine rotor, respectively; $G_{1,1} = \{P_{2,1}, \dot{W}_{ele,1}, T_{2,1}, T_{7,1}\}$, is same as the set as given by (5.4a). We assume

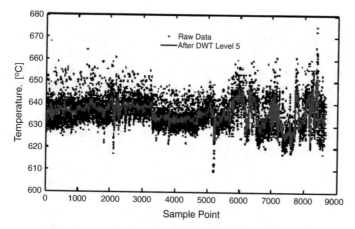

Fig. 6.13 Temperature at the inlet to the 3rd stage of the axial turbine

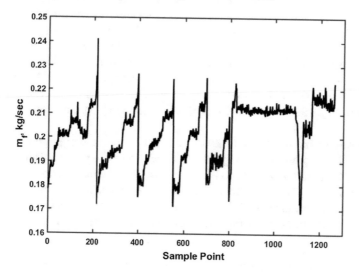

Fig. 6.14 Plots of fuel flow rate data for model training: load less than 50% rated

that VIGV and electric power output are considered to depict the operating region in Euclidian space R^2. Accordingly,

$$\Psi_1 = \{W_{ele,n}, \theta_{IGV}\}$$

In the following, we consider the gas path data and develop models comprising compressor discharge pressure P_2, turbine temperature T_5, and electric power output $\dot{W}_{ele}$. A total of 1263 data points for normal operating conditions are collected with time steps of 20 s. Plots of the input and output data are as shown in Figs. 6.14, 6.15, 6.16 and 6.17.

Less than 50% of rated load region is difficult to represent by models developed from first principles. With the turbine shaft operating at constant speed and the VIGV fully open and remain open until the load passes the 50% boundary, the compressor discharge pressure tends to increase with the increase in the load. The same is true for the temperature at the inlet of the 3rd stage of the gas turbine. The rise in temperature T_5 is acceptable as the fuel flow rate is also increasing. However, the increase in pressure without the change in the airflow rate is difficult to justify for the shaft speed is constant and the VIGV is not manipulated. It may be the case that the bleed valve might be designed to operate in this region as well. But, there is no supporting evidence to uphold the argument. It can be said that, it is quite a weak attempt if meaningless correlations from the point of view of the correlations being liable to be realized in actual controller settings are assumed to link the bleed air quantity with electric power output for the sake of just satisfying the overall system energy balance and mass balance equations. This is basically one of our main reasons behind relying on actual data to develop models for normal operating conditions in this region.

The computer programs for model identification are written in Matlab R2008a. For running the programs, Dell Inspiron Labtop of 3GB RAM, 32bit Operating system and processor of CPU T5850 @2.16GHz is used. The first model developed is a model to characterize the electric power output at the generator terminal as a function of fuel flow rate to the combustion chamber. For normal operating conditions, both the noise ($\tilde{u}(t)$ and $\tilde{y}(t)$) and sensor faults ($f_u(t)$ and $f_y(t)$) are set equal to zero. Besides, we intend to perform the model identification not applying any noise removal by DWT.

The data is first normalized between 0 and 1 using the design point and alarm setting values as presented in Appendix F, Tables F.5 and F.10. We are going to apply LOLIMOT algorithm, Sect. 3.5.1. Hence, before the identification process starts, the user needs to specify the multiplication factor $\triangle$ to define the cluster radius and set of optimization termination criteria. In all the three models that follow, $\triangle$ is assumed 1/3, maximum number of partition is set to 15 and the maximum relative error is set tot 0.01.

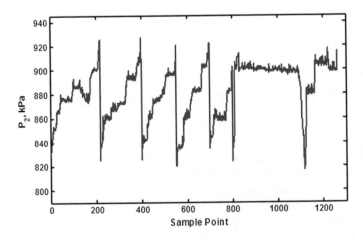

Fig. 6.15 Plot of compressor discharge pressure: load less than 50% rated

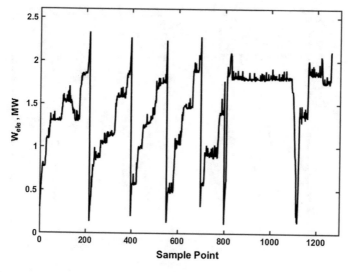

Fig. 6.16   Plot of power output at the generator terminal: load less than 50% rated

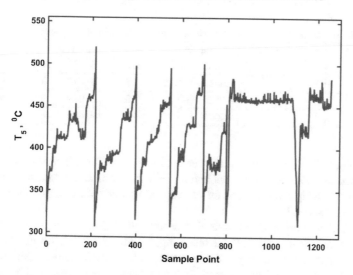

Fig. 6.17   Plot of GT 3rd stage Inlet temperature: load less than 50% rated

After training, the number of rules for P_2, $\dot{W}_{ele}$, and T_5, appeared as 8, 8 and 4, respectively. Figure 6.18 shows the comparison between actual data and predicted result for P_2. In the less than 50% rated load region, the pressure increases with the load until the turbine reaches to the low emission region. Given the model developed based on just six days data, the prediction result compares well with the actual data. The VAF value for both model training and validation are higher than 97%. The rest of the performance parameters calculated for the three models are given on Table 6.1.

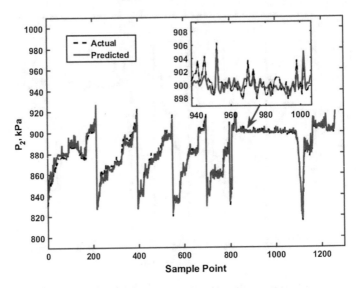

Fig. 6.18 Plot of compressor discharge pressure: load $< 50\%$ rated and $\triangle = 1/3$

Table 6.1 Models performance based on normalized data for training and validation runs

Parameter	Number of rules	Training data			Test data		
		RMSE	AIC	VAF	RMSE	AIC	VAF
P_2	8	0.0027	−10.81	97.7	0.0028	−10.7	97.5
$\dot{W}_{ele}$	8	0.0125	−7.75	98.1	0.0134	−7.6	97.9
T_5	4	0.0091	−8.37	−97.5	0.0095	−8.3	97.3

According to the high values in the performance parameters, it can be said that the developed models are accurate enough for fault detection and diagnosis. Plot of the fuzzy membership functions and the corresponding rule activations levels are shown in Figs. 6.19 and 6.20.

In a further attempt, DWT is used to remove the noises in the input-output data prior to model identification. In the modeling, we have examined the effect of reducing noise levels on the number of rules and the model performance. Table 6.2 summerises the achieved result. It can be seen that the process has resulted in improved performance in terms of number of rules and model accuracy. In the test case, Daubechies wavelets with level-5 are employed with the help of Wavelet Toolbox from Matlab. The effect of the noise removal on the model confidence interval is indicated in Fig. 6.21. The application of DWT reduced the confidence interval. This in fault detection and diagnosis put the model more on the conservative side. With the fuzzy approach in the diagnosis module, this can be a good advantage. Further reduction of the noise is possible but after a certain level, it has the tendency of spoiling the average signal which is critical to model identification. Hence, we limit the noise removal level to a maximum of 5 only.

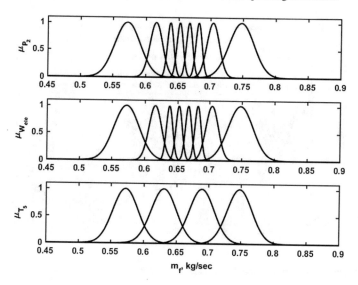

Fig. 6.19 Plots of fuzzy membership functions for P_2, $\dot{W}_{ele}$, and T_5

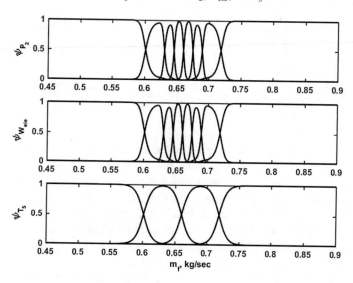

Fig. 6.20 Plots of rule activation levels for P_2, $\dot{W}_{ele}$, and T_5

Table 6.2 Models performance based on normalized data for training and validation runs: noise removed by DWT with db3 and level-5

Parameter	Number of rules	Training data			Test data		
		RMSE	AIC	VAF	RMSE	AIC	VAF
p_2	6	0.0023	−11.15	98.71	0.0023	−11.15	98.15
$\dot{W}_{ele}$	4	0.0105	−8.09	98.48	0.0105	−8.09	98.49
T_5	4	0.0068	−8.95	98.39	0.0068	−8.95	98.40

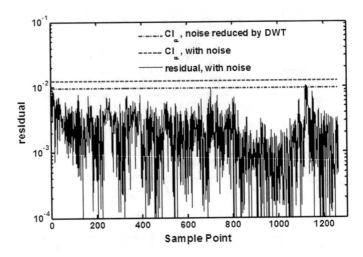

Fig. 6.21 Plots of residual and confidence intervals for P_2

Table 6.3 Number of principal components (PCs) and CumVar for GTG model

Model	Number of variables	Number of PC's	%CumVar$_j$
Gas path	13	5	98.85
Lube system	6	2	99.7
Generator coils	3	1	99.44
T_5 sensors and ΔP_{sd}	7	2	99.77

PCA models are also developed using the training data available for less than 50% load region. For the PCA models, the data are normalized with respect to mean and standard deviation (refer to the equations stated in Appendix C. The loading vectors and their corresponding eigenvalues are obtained by solving an eigenvalue decomposition of the covariance matrices. For deciding on the optimal number of retained loading vectors, Appendix C, Cumulative Variance (CumVar) estimation [6] technique is adopted. The CumVar calculated for the principal components corresponding to the four models are shown in Table 6.3.

For gas path measurements, the first five principal components explain 98.85% of total variance of the system. For the lube oil system and T_5 sensors, the cumulative variance is higher than 99% if two principal components are maintained for each model. For the third model that corresponds to the generator winding temperature sensors, only one principal component is enough to explain 99.44% of the data variance. Assuming a confidence level of 95% for each model, the threshold values for Q-statistic are found to be 0.5292, 0.0644, 0.0649 and 0.0474, respectively. Figure 6.22 shows the SPE or Q-statistic plots for the four models corresponding to the test data. For most of the samples the Q-statistic are below the threshold values. This demonstrates that the PCA is capable of capturing the correlation among the

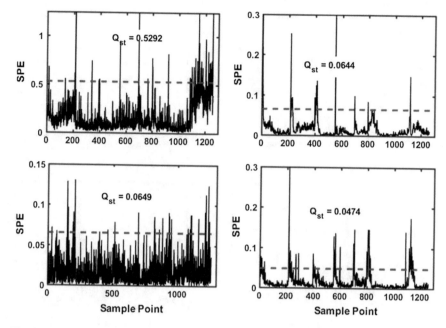

Fig. 6.22 Q-statistic plots for GTG: without DWT

measured variables. If DWT is applied Daubechies wavelets with level-5 the number of retained PCs will remain the same. However, the number of false crossings of the threshold by Q-statistic reduces, Fig. 6.23.

GTG Models for SoLoNOx Operating Region

In the previous section, a working region featured by VIGV fully open and only the fuel flow valve manipulated is considered. In the following the SoLoNOx region is studied. Unlike the low load region, the SoLoNOx region is featured by both the fuel flow valve and VIGV controlled to meet the load and high temperature gas demand. To fully capture the state of the system, a model set that could be called $M_2 = \{\Psi_2, V_2, G_2\}$ need to be formulated. This is according to the concept suggested in Chap. 5 and Sect. 5.4. Four measured gas path parameters vary in this operating region. Accordingly, the subset Ψ_2 can be assumed as $\{W_{ele,n}, \theta_{IGV}\}$, $\{W_{ele,n}, \dot{m}_f\}$, or $\{W_{ele,n}, P_2\}$. For the corresponding fuzzy sets, the same partitioning as given in Chap. 5 and Sect. 5.4 can be adopted. The subset G_2 carries sets of the models in the region. Because the same subsystems and measured signals are available for model identification, it will be considered as $G_2 = \{G_{2,1}, G_{2,2}, G_{2,3}, G_{2,4}\}$ with $G_{2,1}, G_{2,2}, G_{2,3}$ and $G_{2,4}$ tagged to the model sets related to the sensors in the gas path, lubrication system, generator coils, and temperature sensors at the inlet to the 3rd stage of the gas turbine rotor, respectively. Only for the gas path, $G_{2,1}$ is defined as the set of two variables, $G_{2,1} = \{P_{2,1}, \dot{W}_{ele,1}\}$. In the sequel, models in $G_{2,1}$ are analyzed.

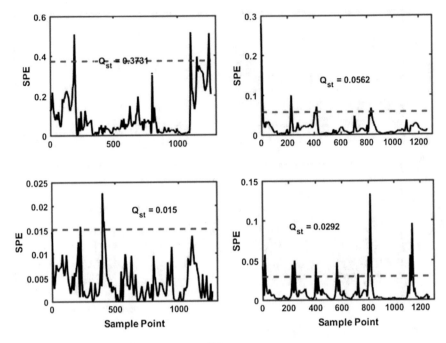

Fig. 6.23 Q-statistic plots for GTG: with DWT

Table 6.4 Models performance based on normalized data for training and validation runs: SoLoNOx region

Parameter	Number of rules	Training data			Test data		
		RMSE	AIC	VAF	RMSE	AIC	VAF
P_2	5	0.0077	−8.93	98.2	0.0084	−8.5	98
$\dot{W}_{ele}$	4	0.0092	−8.35	98.3	0.0093	−8.3	98

We used LOLIMOT algorithm. The constant related to the spread terms is kept 1/3. Six days data total of 28,369 points from isochronous operation is processed for model training and testing. Because both the VIGV position and fuel flow rates are varied, the models for P_2 and $\dot{W}_{ele}$ are designed as functions of the two inputs $\dot{m}_f$ and θ_{IGV}. However, no delay terms are involved in the models. As can be seen from Table 6.4, the developed models featured high accuracy for both the training and test data. The number of rules for each model are also relatively few. For the compressor discharge pressure, the estimated rules are:

$$\text{If } \theta_{IGV} \text{ is } \mu_{11} \text{ and } \dot{m}_f \text{ is } \mu_{12} \text{ then } z_1 = 0.2323 + 0.5966\theta_{IGV} + 0.0993\dot{m}_f$$

$$\text{If } \theta_{IGV} \text{ is } \mu_{21} \text{ and } \dot{m}_f \text{ is } \mu_{22} \text{ then } z_2 = -0.0156 + 0.4326\theta_{IGV} + 0.270\dot{m}_f$$

If θ_{IGV} is μ_{31} and $\dot{m}_f$ is μ_{32} then $z_3 = 0.1507 + 0.4370\theta_{IGV} + 0.3158\dot{m}_f$

If θ_{IGV} is μ_{41} and $\dot{m}_f$ is μ_{42} then $z_4 = 0.2753 + 0.6066\theta_{IGV} + 0.0657\dot{m}_f$

If θ_{IGV} is μ_{51} and $\dot{m}_f$ is μ_{52} then $z_4 = 0.1749 + 0.4087\theta_{IGV} + 0.3019\dot{m}_f$

For the electric power output at the generator terminal:

If θ_{IGV} is μ_{11} and $\dot{m}_f$ is μ_{12} then $z_1 = -0.3549 + 0.4262\theta_{IGV} + 0.9355\dot{m}_f$

If θ_{IGV} is μ_{21} and $\dot{m}_f$ is μ_{22} then $z_2 = -0.5186 - 0.2724\theta_{IGV} + 1.6213\dot{m}_f$

If θ_{IGV} is μ_{31} and $\dot{m}_f$ is μ_{32} then $z_3 = -0.2855 + 0.2471\theta_{IGV} + 0.9567\dot{m}_f$

If θ_{IGV} is μ_{41} and $\dot{m}_f$ is μ_{42} then $z_4 = -0.3101 + 0.0599\theta_{IGV} + 1.1256\dot{m}_f$

Figure 6.24 shows plots of the prediction error and model confidence intervals for P_1 and $\dot{W}_{ele,1}$. As can be observed from the plots, the prediction errors fall within the feasible region defined by 95% confidence level. This is a very good indication that the modes are capable of accurately representing the normal operating condition.

Next, we consider analysis of the performance behind the training algorithms discussed in Chap. 3 Back Propagation (BP), Particle Swarm Optimization (PSO) and Optimum Bounding Ellipsoid (OBE) algorithms. The basic theories behind the

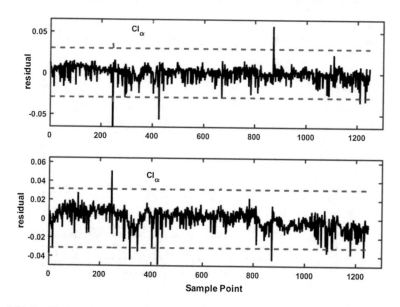

Fig. 6.24 Residuals and confidence intervals for test data: SoLoNOx region

three methods can be found in Sects. 3.5 and 3.6. In applying BP, learning rate and momentum terms of 0.001 and 0.02, respectively, are assumed while a maximum error limit of 0.001 is set to terminate the optimization loop. For PSO algorithm, a similar error limit is considered. But, PSO needs proper values for the inertia term and maximum velocity. To this end, it is assumed that at the start of the optimization process a value of 1.4 for the inertia term is considered appropriate. As the optimization proceeds, it will be modified adapting an exponential variation. As such, equation of the following form is considered to reduce the size of the inertia term as the number of iterations increases. In fact, the equation for the inertia term makes the PSO algorithm unique this thesis.

$$\alpha(i) = \alpha_{max} - exp\left(\frac{-100}{i}\right) \tag{6.2}$$

where, $\alpha_{max} = 1.4$: the starting value of inertia term; i: iteration number. In the BE algorithm, the only parameter that needs to be specified is the model error bound. This value is set equal to 0.01. Training runs were conducted for P_1 and $\dot{W}_{ele}$. In both cases, the OBE algorithm is the fastest to converge, Figs. 6.25 and 6.26. The only drawback is the loss of actual NF model structure in the formation of OBE suitable form. In ANN models, however, the drawback may be tolerated for the ANN is already a non transparent model. As compared to BP, PSO tends to converge with reduced number of iteration. However, the training time is quite large. Summary of the training times is given in Table 6.5.

Comparison of Model Uncertainty as Calculated by Different Approaches

Robust calculation of confidence intervals for normal operating condition is crucial to a successful execution of fault detection and diagnosis in the GTG. In Chap. 4

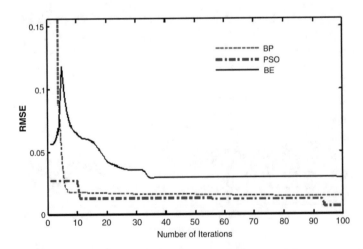

Fig. 6.25 Convergence performance comparison between BP, BE and PSO based algorithms: P_2 model with six rules, $r_c = 0.24$

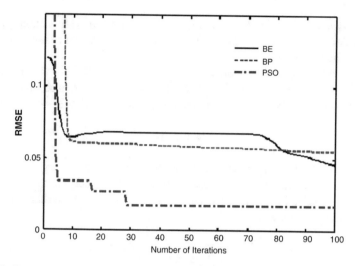

Fig. 6.26 Convergence performance comparison between BP, BE and PSO based algorithms: $\dot{W}_{ele}$ model with six rules, $r_c = 0.24$

Table 6.5 Comparison between performances of different learning algorithms

Variable	Training time (s)				
	BP	PSO	OBE	OB-PSO	JCK
P_2	49.8	2251	9.09		
$\dot{W}_{ele}$	49.9	2038	9.52		

of the thesis, we have presented the models for confidence interval estimation. It is important that the relative difference among the methods be tested before adapting any of them in the FDD system. This section is dedicated to the demonstration of their difference.

If all the model parameters are thought contributing to the model confidence interval, then the equation that we intend to apply is the one given by (4.41). If the assumption is to rely on the parameters in the consequent part of the model only, then (4.21) which is modified form of (4.41) or (4.36) can be applied. In Fig. 6.27 is plotted the confidence intervals estimated for $\dot{W}_{ele}$ model according to the (4.36) and (4.41), respectively. The confidence interval is calculated for a NF model trained by BP algorithm, Sect. 3.5.3. Including all the model parameters in the confidence interval estimation resulted in a relatively conservative feasible region. But, the reduction is not that significant. In fact it can be compensated by assuming a lower confidence level α. Hence, it can be said that truncating the parameters related to the membership functions has minimal effect on the overall confidence interval and reduced form can be used to eliminate any derivative calculation.

NF model training by LOLIMOT algorithm needs less training time as compared to BP and PSO algorithms. In contrast to BE algorithm, it also preserves transparent structure of the NF model. With the abovementioned observations that confidence interval calculations relying only on the consequent parameters are enough to provide accurate description, the use of LOLIMOT algorithm for model training goes quite well with the approximation.

Now, making use of the conclusion from the above analysis, we intend to apply OBE algorithm to estimate model parameters and their confidence limits. As discussed in Sect. 4.2.1, the OBE algorithm requires that model of the system is described by

$$y = \boldsymbol{\Psi}.\boldsymbol{\theta} \tag{6.3}$$

For the FFM-ANN, RBF-ANN and NRBF-ANN models, the regression matrix $\boldsymbol{\Psi}$ can be obtained directly from Chap. 3. For the NF model, the output from the network is described by (3.25). Hence, the use of (6.2) needs further elaboration. Starting with (3.25), we have

$$y_p = \sum_{k=1}^{nk} \varphi_k^p Z_k^p = (\boldsymbol{\varphi}^p)^T \mathbf{z}^p$$

Here,

$$\mathbf{z}^p = [z_1^p \ \dots \ z_K^p \ \dots \ z_{nk}^p]^T$$

$$z_k^p = [1 \ \ \mathbf{u}]\boldsymbol{\theta}_k,$$

and

$$\boldsymbol{\theta}_k = [\theta_{k,0} \ \dots \ \theta_{k,j} \ \dots \ \theta_{k,nu}]^T;$$

Representing $[1 \ \ \mathbf{u}^T]$ by $\tilde{\mathbf{u}}_p$ and $[\boldsymbol{\theta}_1; \ \dots; \ \boldsymbol{\theta}_k; \ \dots; \ \boldsymbol{\theta}_{nk}]$ by $\boldsymbol{\theta}$, the regression matrix resembles:

$$\boldsymbol{\Psi} = \begin{bmatrix} \varphi_1^1 \tilde{\mathbf{u}}_1^T & \cdots & \varphi_k^1 \tilde{\mathbf{u}}_1^T & \cdots & \varphi_{nk}^1 \tilde{\mathbf{u}}_1^T \\ \vdots & \ddots & \vdots & \ddots & \vdots \\ \varphi_1^p \tilde{\mathbf{u}}_p^T & \cdots & \varphi_k^p \tilde{\mathbf{u}}_p^T & \cdots & \varphi_{nk}^p \tilde{\mathbf{u}}_p^T \\ \vdots & \ddots & \vdots & \ddots & \vdots \\ \varphi_1^{Nd} \tilde{\mathbf{u}}_{Nd}^T & \cdots & \varphi_{nk}^p \tilde{\mathbf{u}}_{Nd}^T & \cdots & \varphi_{nk}^{Nd} \tilde{\mathbf{u}}_{Nd}^T \end{bmatrix}$$

where, N_d is the number of data points and $nk4$ is the number of if-then rules in the fuzzy model. After formulating the regression matrix, the model parameter vector $\boldsymbol{\theta}$ can be determined from (6.2) after assuming a suitable error model. Since our intent here is to explore the relative difference in the estimation of model confidence limits, we proceed by considering OBE model training algorithm. We initialize the algorithm assuming $\boldsymbol{\theta}_0 = \mathbf{0}$, $\mathbf{M}_0 = 10^3 \mathbf{I}$ and $\gamma = 0.0001$. Plots of the confidence limits corresponding to $\gamma = 0.0001, \gamma = 0.0005, \gamma = 0.001$ together with the model residual are diagrammed in Fig. 6.28. As can be seen from the figure, the confidence limit relies on the error bound. It gets bigger and bigger as we increase the error bound. As stated in Sect. 4.2.1, the confidence limit is a function of the size of the

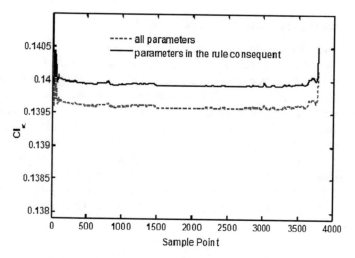

Fig. 6.27 Confidence Intervals for $\dot{W}_{ele}$: NF model trained by BP

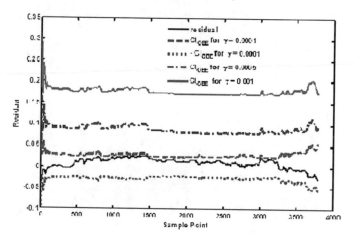

Fig. 6.28 Confidence Intervals for $\dot{W}_{ele}$: NF model trained by OBE

ellipsoid, which in turn is calculated based on the error bound. Hence, the increase of the confidence limit with the error bound is expected. As compared to the confidence limit for 95% confidence level in BP algorithm (see Fig. 6.27), the confidence limit from OBE algorithm becomes higher even for small error bound. The error bound is a function of accuracy of the measuring instrument. Hence, too small values may be too rough to perfectly characterize the instrument. In consideration of the fact that the error bound could be bigger than $\gamma = 0.001$, the use of the OBE algorithm is less than attractive for designing an FDD system sensitive to small system changes. In the thesis, this observation is considered every time when we decide to use iid error assumption.

Performance Related Surface Plots for the GTG

One advantage of a data based model is that the identified model can be used to predict outputs for a different set of inputs on which the model was not trained for. Besides, the calculation time is pretty fast as compared to mechanistic models. In this section, prediction and analysis of performance related parameters covering the whole operating region of the GTG are considered. With the two-dimensional plot of the parameters, it may be difficult to look at the whole possible input-output combinations. As a solution, we have used the static models developed in the preceding sections to carry out 3D surface construction. The first set of plots provide the distribution of compressor discharge pressure P_2, electric power output $\dot{W}_{ele}$, and temperature T_5 as a function of the VIGV position θ_{IGV} and fuel flow rate $\dot{m}_f$, both normalized with respect to the values at the design point, Fig. 6.29. It can be seen

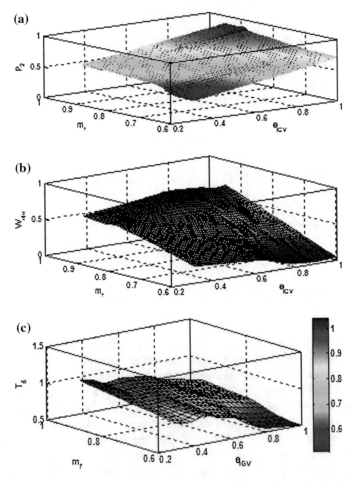

Fig. 6.29 Surface Plots of GTG output parameters: LOLIMOT algorithm with six rules and $\Delta = 1/3$: **a** P_2, **b** $\dot{W}_{ele}$, and T_5

from the graphs that P_2 is maximum for higher values of VIGV opening and fuel flow rate. The electric power output seems maximum at higher fuel flow rate, and VIGV opening in the range of 0.6–0.8. In a similar analysis, the temperature at the inlet of the third stage of the turbine T_5 is lower for VIGV near to fully open position and the fuel flow rate relatively low. However, when the VIGV opening is reduced and the fuel flow rate is increased, the temperature also increases.

The other performance parameters included in this section are thermal efficiency η_{th}, Specific Fuel Consumption (SFC), and exergetic efficiency η_{exg}. The equation for each of them is available in Appendix E. Thermal efficiency for the GTG is the ratio between power output at the generator terminal and the energy input to the system. Or, if the Heat Rate (HR) is know then thermal efficiency is the reciprocal of HR. In Fig. 6.30a, surface plot of η_{th} in a non-dimensional space is shown. For θ_{IGV} in the range of 0.6–0.8 and the fuel flow valve almost at fully open setting the thermal efficiency is higher. In fact, the distribution of the thermal efficiency is similar to the distribution of $\dot{W}_{ele}$. The second performance parameter, SFC, is defined as the ratio between the amount of fuel admitted into the combustion chamber and the electric power output. Lower values indicate better performance. SFC is often used to compare performance of different size engines. In Fig. 6.30b, the SFC is demonstrated as a function of θ_{IGV} and $\dot{m}_f$. In the region where θ_{IGV} is greater than one and $\dot{m}_f$ is less than 0.65, the value of SFC gets higher. As indicated in the previous graphs, this happens at a region where both P_2 and $\dot{W}_{ele}$ are a minimum.

The third parameter mentioned to characterize the system is exergetic efficiency. Exergy is the standard measure of energy quality. The maximum work that can be extracted from heat and electric power are different. To this end, the use of exergy is convenient for it accommodates the energy quality difference. Exergetic efficiency η_{th} for the GTG is defined as the electric power output divided by the input energy corrected for quality, Appendix E. Figure 6.30c shows surface plot of the exergetic efficiency. Except the magnitude difference, the plot is similar to the plot for the thermal efficiency. In conclusion, making use of the NF models we have managed to easily generate surface, which would have been time consuming had the first model be applied. The surface plots helps to see how the system behaves for different input parameter combinations. The plots are valid for inlet temperatures in the range of 24–34 °C.

Models for GTG Start-Up Region

A brief description of the start-up system has been given in Sect. 6.2.1. While the start-up includes ramping to 20% speed, purging and running to the synchronous speed, only the third stage is featured by changes in most of the measured parameters. For the starter motor, the input current i_s and voltage V_s, and shaft speed Ω_{st} are measurable. The starter motor provides torque to the GTG main shaft until the GTG reaches the self-sustaining speed (65% synchronous speed). Once it reaches to that speed, the starter motor disengages itself automatically. The lube system and generator coil temperatures are also available measurement. The models presented in this section encompass those related to the gas path system and starter motor only. Since the GTG speed is dependent on the VIGV position θ_{VIGV}, fuel flow rate $\dot{m}_f$ and starter

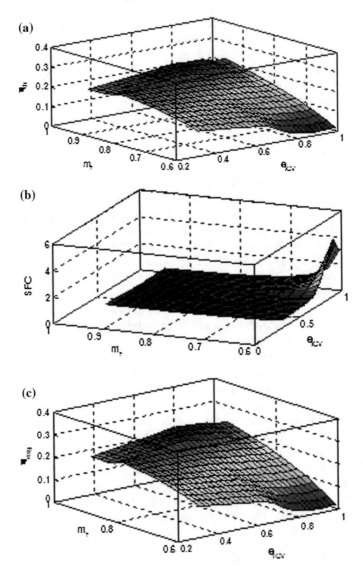

Fig. 6.30 Surface Plots of GTG performance parameters: LOLIMOT algorithm with six rules and $\Delta = 1/3$: **a** η_{th}, **b** SFC, and **c** η_{exg}

shaft speed, the three inputs are taken into account while forming a model for the synchronous speed. The four models in the start-up region are defined by (6.3)–(6.6). LOLIMOT algorithm with $\Delta = 1/3$ is selected for model training. The training is performed based on one week data while another data after one week later is used for model testing.

Table 6.6 Models performance based on normalized data for training and validation runs: start-up region

Parameter	Number of rules	Training data			Test data		
		RMSE	AIC	VAF	RMSE	AIC	VAF
Ω_{sh}	14	0.0031	−10.1	99.97	0.0082	−7.5	99.8
P_2	18	0.0077	−8.3	99.78	0.0228	−5.5	97.9
T_5	18	0.0171	−6.7	97.9	0.0873	−3	53.77
Ω_{st}	18	0.313	−5.5	98.7	0.1435	−2	73.85

$$\Omega_{sh} = f(\theta_{VIGV}, \dot{m}_f, \Omega_f) \tag{6.4}$$

$$P_2 = f(\theta_{VIGV}, \Omega_f) \tag{6.5}$$

$$T_5 = f(\theta_{VIGV}, \dot{m}_f) \tag{6.6}$$

$$\Omega_{st} = f(i_s, V_s) \tag{6.7}$$

Model performance parameters corresponding to training and testing runs are listed in Table 6.6. As can be observed, the models demonstrate higher performances for the training data while lightly reduced values are experienced for the testing data. The reduced performances are acceptable for they are small and within in the confidence limits for normal operating regions. Four less rules are included in the model for the starter motor shaft speed. This is due to the trend in the normal operation data.

Plots of predicted results for the four parameters are portrayed in Fig. 6.31. For GTG shaft speed higher than 65% of the synchronous speed, the prediction and actual data match very well. Once the purging stage is passed, the starter motor shaft speed increases for a while until the GTG speed reaches 65% speed and then decreases to zero. The temperature T_5 is higher at around 56% speed for the combustion is already started and the VIGV is yet to open gradually. As VIGV opening continues, the temperature starts decreasing accordingly. This is visible in the prediction plots. The discrepancy between the prediction and actual data in the less than 65% speed region can be attributed to availability of limited data and outliers in the measurement. As such, the models can be improved by considering more data for model training. As expected, the compressor discharge pressure increases with increase in GTG speed. In light of the high values in the performance parameters (see Table 6.6), and the good agreement in greater than 65% speed region, it can be concluded that the models are accurate enough for capturing normal operating condition.

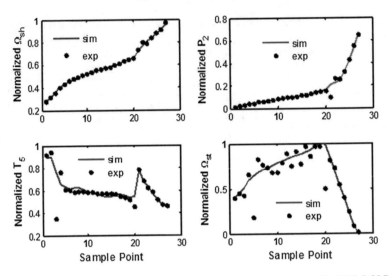

Fig. 6.31 Comparison between actual and predicted data in the start-up region for GTG: LOLIMOT algorithm with $\Delta = 1/3$

Identification of GTG Dynamic Models

In this section, we present the dynamic models developed for the gas turbine generator. Unlike steady state models, the dynamic models involve delayed input and output signals. As mentioned in Sects. 3.2.1 and 5.6, the output delay term or the autoregressive part can be eliminated by developing the nonlinear model in the frame work of OBFs. Hence, the dynamic models presented hereunder are all incorporating OBF unless exceptions are made for model comparison cause. Another point worth noting is a dynamic model may refer to less than 50% rated load region or SoLoNOx region. To avoid repetitive application of the proposed concept to the same system, the analysis focuses on the second operating region for the whole purpose of this section is to investigate the effect of different values of the dominant pole and model order on the performance of the OBF based NF model.

The first question we ask is if the pole can be assumed arbitrarily. It was shown in [7, 8], with an example on Magnetic Levitation, that the choice of ξ is not so important. However, if the pole size is selected close to the dominant pole of the system, a better representation for a given model order may be possible. For the GTG, trying to get the dominant pole from mechanistic models, as the others did, is difficult because of the reasons partly mentioned in Sect. 6.2.1. In the research work by [9], a two step optimization procedure is suggested. First NF model is developed assuming ARX local model structure. Then, time constant τ of the system is estimated from step response of the model. Based on the time constant, the dominant poles is approximated as $\xi = exp(= T_s/\tau)$. Where, T_s is the time step.

For the GTG, it is more justifiable than otherwise if we assume that the system is a well damped system. As stated in Sect. 3.2.1 and latter repeated in Sect. 5.5, for a

Table 6.7   Effect of dominant pole ξ on the Performance of NF-OBF model: $\dot{W}_{ele}$

Pole ξ	n_k	Training performance				Testing performance		
		RMSE	AIC	VAF	Time	RMSE	AIC	VAF
0.5	7	0.0165	−7.19	94.7	82.4	0.0173	−7.1	94.2
0.6	8	0.0164	−7.2	94.8	106	0.0179	−7.03	93.8
0.7	8	0.0167	−7.17	94.63	104	0.017	−7.13	94.4
0.8	9	0.0168	−7.15	94.5	136	0.02	−6.8	92.3
0.9	7	0.0174	−7.1	94.1	79.6	0.183	−6.98	93.5
0.95	8	0.0185	−6.96	93.4	106	0.0186	−6.94	93.2

Table 6.8   Effect of Model Order on OBF-NF Model Performance, $\xi = 0.7 : \dot{W}_{ele}$

Model Order, m ξ	n_k	Training performance				Testing Performance		
		RMSE	AIC	VAF	Time	RMSE	AIC	VAF
2	8	0.0173	−7.1	94.2	54	0.0173	−7.1	94.2
4	8	0.0167	7.17	94.63	105	0.017	−7.13	94.36
6	10	0.0164	−7.19	94.8	264	0.0168	−7.14	94.5
8	10	0.0155	−7.3	95.33	364	0.0237	−6.44	89.03

well damped system, Laguerre basis functions serve very well. In the following, we assume a single real pole $\|\xi\| < 1$ to define the Laguerre bases.

Table 6.7 indicates model performance parameters calculated for different values of ξ and corresponding to $\dot{W}_{ele}$. For a fixed number of OBFs, variation in pole size seems having minimal effect on the model performances. Even the number of rules generated by LOLIMOT algorithm is less affected by the pole size. It can be said that, the assumption of an arbitrary pole works satisfactorily if the model training is performed by LOLIMOT algorithm. LOLIMOT algorithm adds rules until modeling error reaches a preset error limit (see Chap. 3 and Sect. 3.5.2).

The other test that we have conducted is the analysis on the effect of number of OBFs with fixed pole on the model performances. Here as well the same learning algorithm is relied up on. For linear systems, a good choice on the dominant pole leads to few number of OBFs or model order. As can be seen in Table 6.8 an increase in number of OBFs tends to slightly improve the model performance but with more fuzzy rules involved and the training time also increased. Validation plot for P_2 and $\dot{W}_{ele}$ are displayed in Figs. 6.32 and 6.33, respectively. In each case, the number of OBFs and pole size are selected as m and $\xi = 0.7$, respectively. Both NF-OBF modes have demonstrated high accuracy.

Validation Result for Approximate Thermodynamic Model of the GTG in the Loaded Region

In this part of the thesis, we focus on demonstrating validation of the approximate thermodynamic model developed for the GTG. As mentioned in Sect. 6.2.1, actual simulation of a GTG needs performance maps for the component parts which are

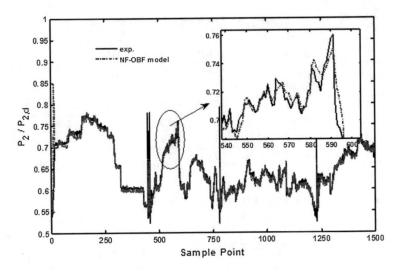

Fig. 6.32 NF-OBF model for P_2: $m = 4$ and $\xi = 0.7$

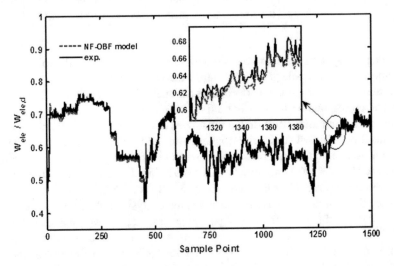

Fig. 6.33 NF-OBF model for $\dot{W}_{ele}$: $m = 4$ and $\xi = 0.7$

hardly available. Presented is a semi-empirical approach in which the use of compressor map is avoided by adapting regression models to approximate the compressor in the whole operating region. For the turbine, the traditional scaling method, Appendix E, is used to generate representative characteristic curves for the turbine stages. The design data is obtained from overall characteristic map of the GTG, which is commonly supplied by the manufacturer. The rest of the calculations follow the common design and off-design calculation procedures well documented in Saravanamutto et al. [1], Walsh and Fletcher [2], and Razak [4]. The relevant equations and the calculation algorithms are summarized in Appendix E while the algorithms for design point and off-design point calculation are documented in Appendix D. The models used for the axial compressor are the following. After curve fitting the actual data, the normalized compressor discharge pressure is approximated by (6.7) and (6.8).

$$\left(\frac{P_2}{P_{2,d}}\right) = \begin{cases} 0.1955 \left(\frac{\dot{W}_{ele}}{\dot{W}_{ele,d}}\right) + 0.7969, \ for \ \left(\frac{\dot{W}_{ele}}{\dot{W}_{ele,d}}\right) \leq 0.5 \ and \ \theta_{VIGV} = 100 \\ 0.7114\theta_{VIGV} + 0.3479, \ for \ \left(\frac{\dot{W}_{ele}}{\dot{W}_{ele,d}}\right) > 0.5 \end{cases}$$

$$(6.8)$$

And, for the compressor efficiency:

$$\left(\frac{\eta_{AC}}{\eta_{AC,d}}\right) = \begin{cases} 1 - b + 11 \left(1 - \frac{P_2}{P_{2,d}}\right)^2 .for \ \left(\frac{\dot{W}_{ele}}{\dot{W}_{ele,d}}\right) \leq 0.5 \ and \ \theta_{VIGV} = 100 \\ 1 - b_{1,2} \left(100 - \theta_{VIGV}\right)^2, \ for \ \left(\frac{\dot{W}_{ele}}{\dot{W}_{ele,d}}\right) > 0/55 \end{cases}$$

$$(6.9)$$

Where, $P_{1,d}$, $\eta_{AC,d}$ and $\dot{W}_{ele}$ are design point values of compressor discharge pressure, isentropic efficiency, and electrical power output at the generator terminal, respectively. The calculated and read values of each of these parameters are given in Appendix F, and Tables F.4 and F.5. The constants $b_{1,1}$ and $b_{1,2}$ are set to be 0.5 and 8e-5, respectively. The use of similar set of equations can be found in the work of Haglind [10]. The difference in our case is that the GTG is not a new one and it was necessary to approximate the current condition in the stated way so that already reduced performance can be accounted for. Latter in the testing of the proposed FDD system, test data on process faults can be generated using the same models. Equations (6.7) and (6.8) work for compressor inlet temperature in the range of 24–34 °C. For the turbine, considering the widely used data from Converse [11], the resulting generalized performance maps after scaling are as shown in Figs. 6.34 and 6.35. In our model, since the turbine involves blade cooling, a stage by stage approach

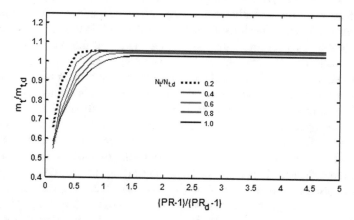

Fig. 6.34 Generalized mass flow rate versus pressure ratio

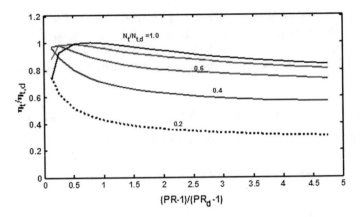

Fig. 6.35 Generalized mass flow rate versus pressure ratio

adapting the generalized performance maps to each stage is employed. The mass and energy conservation equations for each turbine stage are discussed in Appendix E.

In the model of the GTG, pressure losses in the suction duct, compressor diffuser, and combustion chamber are all included. This is in addition to the models introduced to account for the effect of bleed air for blade cooling. During simulation, the error in the estimation of the two major parameters temperature T_5 and electric power output are minimized. In the sequel, the discussions on the resulting plots are outlined. It is worth mentioning that, the GTG is not installed for conducting an experiment. Due to this, it was not possible to keep the inlet temperature constant and vary the electricity and heat load. Therefore, what is presented is based on actual operation data in which the load demand varies with the atmospheric temperature. Gearbox and generator efficiencies are described based on the models given in [10]. Summary of the models can be found in Appendix E.

Figure 6.36 presents the first result. It demonstrates the variation of normalized compressor discharge pressure, temperature T_5 and fuel flow rate as a function of relative load. In the first operating region for relative load less than 0.5 the temperature T_5 increases with load. This is expected because the GTG in this region is operating in load control only. To meet the load requirement, the fuel flow rate increases gradually with the load. During this time, the compressor VIGV is at fully open position and the VSVs are all at their respective design stagger angles that make the air flow rate almost constant. The normalized pressure also increases with lower slope. This is quite strange for a compressor whose shaft speed is constant and VIGV at fully open position. One may argue that this may be due to the existence of bleed air from the compressor. In our case, there is no strong evidence that supports the use of bleed air during part load. The document provided by the manufacturer only mentions the use of the bleed valve during starting and shut-down. After many simulation tests, excluding bleed air during part load operation, the algorithm had no troubles in convergence if the actual pressure ratio according to (6.7) is used to match the temperature T_5 and electric power output. The resulting air mass flow

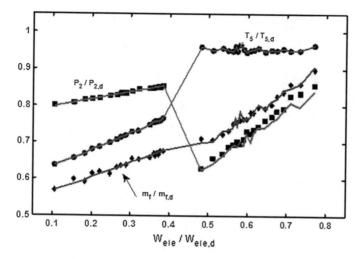

Fig. 6.36 Part Load Performance of the GTG (lines: model, marks: actual data); normalized GTG parameters: compressor discharge pressure, fuel flow rate, and temperature T_5

rate in this region also appears to remain almost constant. In light of satisfying the four conditions, it can be said that the assumption on no bleed air during low load operation seems convincing.

For load greater than 50% of rated load, the GTG runs under temperature and load control. In fact this region is called low green house gas emission region or SoLoNOx region. Near to 50% load, the VIGV starts to close gradually while the setting for temperature T_5 is increased to 667 °C. This is visible in the validation graphs, Fig. 6.36. The monitored data is collected every 10 s. Because the rate of opening of the VIGV and ramping up of T_5 setting in the transition region is faster than 10 s, it was not possible to acquire recoded data for validation purpose. The temperature control is based on the feedback signal from temperature at the inlet to the 3rd stage of the gas turbine. In this region, the hot exhaust gas is used to run the heat recovery steam generator. The same region is operating with low NO_x and CO_2 emissions. This happened so by using a lean premix combustion process. In Fig. 6.37 is shown the temperature predictions at different location of the GTG. As indicated, in the higher load region the temperature T_5 is held almost constant while it is varying elsewhere. The non measurable temperatures T_2, T_3, T_4 and T_7 are successfully predicted over the whole operating region.

The Specific Fuel Consumption (SFC) decreases with increase in rated power while the thermal efficiency increases, Fig. 6.38. These parameters are commonly used to make a comparison between performance of new and used turbines. The SFC is the fuel consumption per unit power output. Hence, the decrease with rated power in Fig. 6.38 is reasonable. The thermal efficiency is the ratio between the power output and the input energy. The expected trend that is increasing with the rated power is visible.

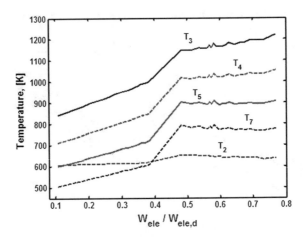

Fig. 6.37 Predicted part load temperatures for the GTG

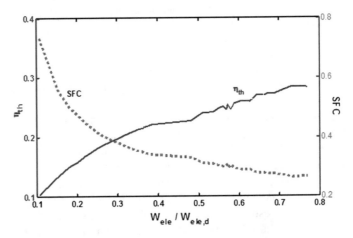

Fig. 6.38 Predicted performance of the GTG: thermal efficiency and specific fuel consumption

In the GTG, one of the parameters that is not measured but very critical to the performance analysis of the system is the air mass flow rate. Basically, air flow measurement is often taken indirectly by installing a venture or orifice and measuring the pressure difference across the venture or orifice. It is often the case that such flow measurement devices are not available in real turbines unless it is in a test rig. Hence, it was necessary to rely on an appropriate thermodynamic model. In Fig. 6.39 is shown the variation of the predicted air mass flow rate with the rated load. As indicated, it remains constant for less than 50.

Equally important is the emission trend over the whole operating region. Normally, the emission of CO is high at low temperatures while it decreases with temperature. The NOx emission, however, is a minimum at low temperatures and increases with increase in temperature. In the GTG, a SoLoNOx system is used to balance the two emissions by using dry low emission combustion. What is plotted in Fig. 6.39 is

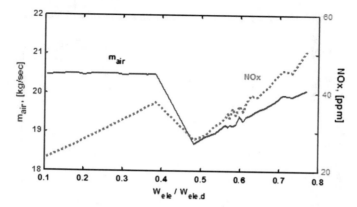

Fig. 6.39 Predicted part load performance of the GTG: mass flow rate of air and NOx emission

the NOx emission as predicted by a model discussed in Razak [4]. Summary of the empirical equation is given in Appendix E. As shown in Fig. 6.39, the NOx emission increases with the rated load. Abnormal condition in the GTG may manifest itself as changes in the NOx trend. This is helpful in the diagnosis of faults. Abnormal condition can also be manifested as derangements in stream exergies or exergy destruction ratio across system components. The stream exergy at the inlet or outlet of a system is a function of inlet temperature, pressure, specific heat and properties at the reference state, Appendix E and (E.5). The exergy destruction ratio for a given unit is calculated as the exergy change across the unit divided by the total exergy destruction in the system. Both quantities are good indicators of increased losses in the system. In fact, they are often used to identify areas where improvements can be made. The equations behind exergy destruction ratio and corresponding to each component of the GTG are described in Appendix E. In Fig. 6.40, we see plots of exergy destruction ratio for each component in the GTG. As can be seen from the plots, for loads greater than 30% of the rated value, the highest exergy destruction occurs in the exhaust duct. This is valid for the GTG has to provide hot gas to the HRSG. It remains constant for higher loads.

The next highest exergy destruction is demonstrated in the combustion chamber. The source of exergy destruction in the combustor is the irreversibility in the combustion process. In the SoLoNOx region, it tends to increase with the load. In Fig. 6.41 are shown stream exergies and exergy change across the combustion chamber. These values are the inputs used for the construction of exergy destruction ratio.

In summary, this section demonstrated validation of the semi-empirical model developed for the GTG. It was shown that the developed model featured high accuracy in the prediction of fuel flow rate. With the model, we were able to quantify non-measurable quantities over the whole operating region. This is quite important for latter design and testing of the proposed FDD system. Simulation of the HRSG model cannot be done without knowing what is supplied by the GTG. In this regard, the GTG model will also be used to simulate the HRSG model for different loading of the GTG.

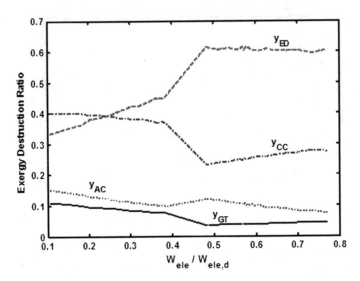

Fig. 6.40 Predicted exergy destruction in each component of the GTG

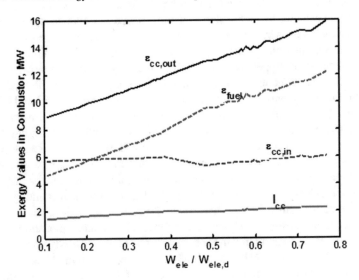

Fig. 6.41 Exergy of streams and exergy destruction in the combustion chamber

6.3.2   Identified Models for the HRSG and SAC

The proposed FDD structure for the HRSG and SAC requires that for normal operating conditions each of the measured variables must be represented by a suitable model. Latter in the procedure, the normal model will be used to predict the output for newly collected data. For the HRSG alone there are twelve measurable variables, Appendix F and Table F.8, and hence there needs to be twelve models. Irrespective of the method used for modeling, it is a common practice to confirm suitability of a designed model through validation by an experimental data. Hence, we do model validation for the two systems.

As stated in Sect. 6.2.2 of the chapter, a first principle model cannot be considered as a preferred method for the design data are mostly not available. In our case the adopted method is NF modeling approach. However, the decision on the number and type of inputs considered for a given output is partially decided by referring to equations formulated according to first principles. For instance, the water level in the steam drum is a function of steam flow rate, heat transfer across the evaporator, feed water inlet temperature, and feed water flow rate. The governing equation also relies on constants related to the working fluid and heat capacity of the physical system. Often, the property equations are arranged as a function of the drum pressure. In the sequel, models for three of the measured variables drum water level, feed water temperature at the economizer outlet, and temperature of exhaust gas at the inlet to the economizer are presented. The decision on the three variables is because of the fact that the inclusion of the whole variables may spoil presentation of the results.

The first result presented is the model developed for the water level in the steam drum, Fig. 6.42. For higher duty, the water level is controlled at about 60%. This is visible in the developed model. To accurately represent the system, the TSK method trained by the LOLIMOT algorithm is adopted. For reduced root mean squared error, the algorithm settled to 10 rules.

The second model considered for the HRSG is the feed water temperature at the economizer outlet. The economizer is used to recover the heat from the exhaust gas. It is also crucial to keep the approach temperature in the acceptable range (5–10 °C). Too high approach temperature causes steaming in the pipe lines connecting the economizer and the steam drum. As can be seen from Fig. 6.43, the feed water temperature is well below the average temperature in the steam drum (T_{sat}

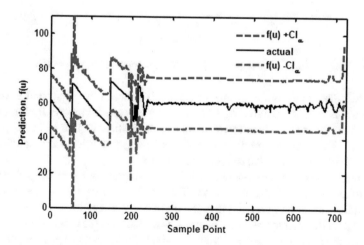

Fig. 6.42 NF model of drum water level in the HRSG, H_D: $\triangle = 1/3$

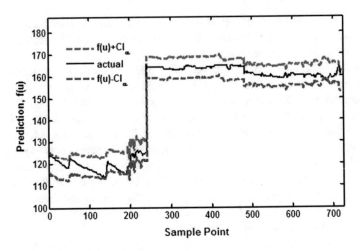

Fig. 6.43 Feed water temperature at the economizer outlet, T_{fd3}: $\triangle = 1/3$

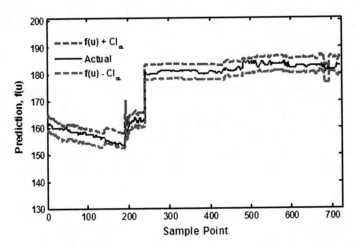

Fig. 6.44 Temperature of exhaust gas at the inlet to the economizer, T_{g2}: $\triangle = 1/3$

$\sim 172\,^{\circ}$C). The trend for the normal operation is between the upper and lower prediction limits that are featured by 95% confidence level. It is worth noting that the Confidence Interval (CI) is estimated applying the method explained in Chap. 4 and Sect. 4.2.1. The resulting prediction model is having five rules corresponding to a root mean square error of 0.0102 for training data.

For normal and efficient operation, it is not only the approach temperature that needs to be in the acceptable region. The pinch temperature — the difference between the saturation temperature at the drum pressure and the temperature of the exhaust gas at the inlet to the economizer — must also be in the range of 5–15 °C [12]. A comparison between actual and predicted trends for the exhaust gas temperature at the inlet to the economizer is shown in Fig. 6.44. Akin to the previous case, we used

Table 6.9 Number of rules (nk) and models performance parameters for the HRSG

Output parameters	nk	Model performance parameters		
		RMSE	AIC	VAF
H_D	14	0.0257	−7.2462	65.915
T_{g2}	7	0.0049	−10.61	99.19
T_{fd3}	5	0.012	−9.14	99.21

a confidence level of $\alpha = 0.5$ that corresponds to 95% confidence limits. As shown in the figure the actual data falls between the upper and lower confidence bounds, which is an indication that the data used for validation is for normal operation. The number of rules and model performances can be seen in Table 6.9.

Considering the SAC, about eight parameters are available for measurement, Appendix F and Table F.9. The load in the SAC is indicated by the chilled water temperature at the inlet to the evaporator (T_{ch1}). As mentioned in Sect. 6.2.3, temperature of the chilled water leaving the evaporator is controlled to about 6.5 °C regardless of the inlet temperature. Hence, the trend of T_{ch1} is a good indicator of the load in the SAC. In the sequel, we intend to consider models for T_{ch1}, cooling water outlet temperature (T_{co2}) and steam flow rate ($\dot{m}_f$). The three parameters are important to demonstrate how the system behaves. Latter in the presentation of the proposed FDD system, the models will be used to help verify the presence of a specific type of fault in the system. On top of that, the accurate modeling of the two parameters may lead to better estimation of performance indicators like Coefficient of Performance (CP) and exergetic efficiency (η_{exg}).

For T_{ch1}, actual data and its validation trends are both plotted in Fig. 6.45. Model training is performed based on LOLIMOT algorithm. The temperature is

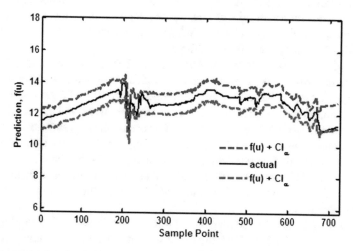

Fig. 6.45 Comparison between actual and predicted values for (T_{ch1}) in °C

Table 6.10 Number of rules (nk) and models performance Parameters for the SAC

Output parameters	nk	Model performance parameters		
		RMSE	AIC	VAF
T_{ch1}	10	0.0163	−8.1698	88.50
T_{ch2}	8	0.0642	−5.4581	76.01
$\dot{m}_s$	10	0.8328	−0.3105	90.67

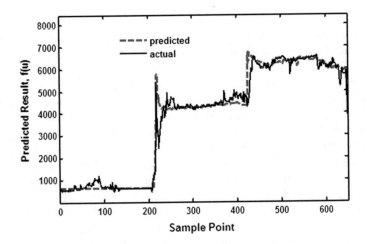

Fig. 6.46 Comparison between actual and predicted values for ($\dot{m}_s$) in kg/s

thermodynamically related chilled water valve position (θ_{vp}), T_{ch2}, $\dot{m}_s$ and T_{co1}. Due to this reason, data for the four parameters are used as an input to develop a model for T_{ch1}. The steam flow rate to the SAC is varied based on the magnitude of the chilled water inlet temperature. In light of that, the development of a model for taking into account the steam flow rate is logical. Summary of the model performance in terms of RMSE, AIC and VAF are listed in Table 6.10.

Application of 1st-law of thermodynamics and heat transfer equations confirms that the eight measurable parameters for the SAC are correlated. Making use of the energy equation for the SAC, a model is developed for the steam flow rate as a function of T_{ch1}, T_{ch2} and T_{c01}. The prediction from the resulting model and the actual data are diagrammed in Fig. 6.46. The model, while having ten rules, has sufficiently matched the actual data. The steam is the source of driving energy for the SAC. Hence the increase of the steam flow rate with the increase in the chilled water temperature is something expected. Accuracy of the predicted result is convincing enough for FDD design as it falls in the region that is bounded by the confidence limits for normal operating conditions.

The third model considered for analysis refers to temperature of the cooling water leaving the SAC. The heat from the absorber and condenser are removed by the cooling water. It increases with the SAC load. It is also a good indicator of the health

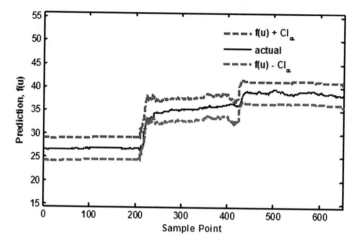

Fig. 6.47 Comparison between actual and predicted values for (T_{co2}) in °C

state of the cooling system. From the energy equation, the outlet temperature is related to steam mass flow rate and inlet temperature T_{co1}. In Fig. 6.47 is shown the actual data and the confidence limits for normal operating conditions. The confidence limits are the result of modeling the temperature by NF structure having eight rules. The actual data is between the maximum and minimum predictions. Hence, it can be said that the model can be considered in the design of temperature detector. Calculated performance parameters are depicted in Table 6.10.

6.3.3 Validation of First Principle Model for the HRSG

In this section is discussed validation of the approximate thermodynamic model developed for the HRSG. The governing equations are elaborated in Appendices E.10 and E.11. Since the pinch point and approach T_{ap} temperatures T_{app} and T_{ap} are not known at the design point, iterated values in the recommended regions [12] are selected. Calculated values according to the algorithm given in Fig. D.7 are $T_{pp} = 22°C$ and $T_{ap} = 11°C$, respectively. Summary of calculated design point data are outlined in Table 6.11. The overall heat transfer coefficient for the evaporator is the highest in the system. Hence, we expect longer width and height. Exhaust gas temperature T_{g2} and feed water temperature at the inlet to the economizer T_{fd3} are two of the parameters measured for condition monitoring. These temperatures are considered for model validation. Figure 6.48 shows the comparison between actual and predicted value for the exhaust gas temperature T_{g2} as calculated by the algorithm whose flow chart is shown in Fig. D.8. In the off-design calculation, a continuous blow-down of 6% is assumed regardless of the duty. The design point heat transfer coefficients are tuned to match the actual trend. For the evaporator, the

Table 6.11 Summary of calculated design point data for the HRSG

Parameter	Symbol	Unit	Value
Pinch point temperature	T_{pp}	°C	22
Approach temperature	T_{pp}	°C	11
Evaporator heat transfer coefficient	UA_{evp}	W/K	61
Economizer heat transfer coefficient	UA_{evp}	W/K	29.04
BD heat exchanger transfer coefficient	UA_{evp}	W/K	1.76
Evaporator tuning parameter	K_{evp}	1	0.02
Economizer tuning parameter	K_{evp}	1	−0.2

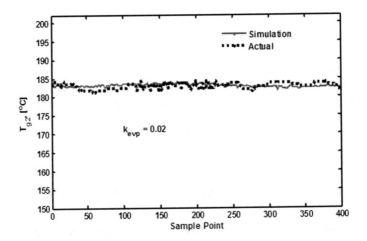

Fig. 6.48 Comparison between actual and predicted values for (T_{g2})

tuning parameter K_{evp} is set equal to 0.02 while for economizer, k_{eco} is determined to be −0.2. As shown in Figs. 6.48 and 6.49, the predicted graphs are closer to the actual data. The average relative prediction error for T_{g2} and T_{fd3}, respectively, are 0.4246 and 1.8776%. In light of these and given the limited available design information, it can be said that the model is good enough for generating fault detection and diagnosis relevant data.

Having a thermodynamic model also allows estimating parameters that are commonly not available for measurement. In Fig. 6.50 are plotted the variation of pinch temperature (T_{pp}) and approach temperature (T_{ap}). The temperatures are in the recommended regions. The approach temperature has the tendency of varying with temperature of the exhaust gas. Figure 6.51 shows the variation of exergetic

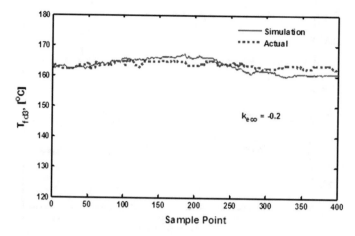

Fig. 6.49 Comparison between actual and predicted values for (T_{fd3})

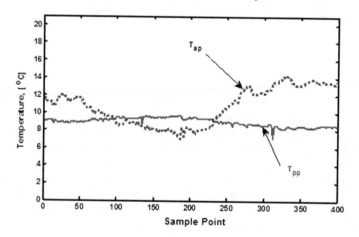

Fig. 6.50 Predicted pinch (T_{pp}) and approach (T_{ap}) temperature

efficiencies for the evaporator, economizer and blow-down heat exchanger. Exergetic efficiency of the economizer is higher than that of evaporator and blow-down heat exchanger. One reason that can be mentioned is the temperature of the exhaust gas at the inlet to the economizer is relatively lower and most of the energy is recovered. The exergetic efficiency is a good indicator for areas where improvements could be made to increase the overall efficiency. Based on the plots, the blow-down heat exchanger seems a good candidate for improvement.

The developed thermodynamic model has demonstrated reduced relative percentage prediction error. Besides, it allowed estimating fault detection and diagnosis relevant parameters, which are often not available for measurement. Hence, we conclude by saying that this model will be used in Chap. 7 to validate the proposed FDD system.

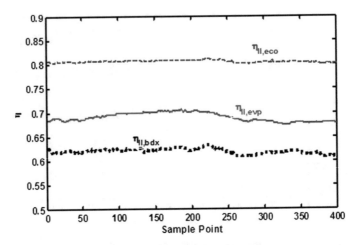

Fig. 6.51 Exergetic efficiencies for the evaporator ($\eta_{II,evp}$), economizer($\eta_{II,eco}$), and blow-down heat exchanger ($\eta_{II,bdx}$)

6.4 Summary

The aim of this chapter has been to present model identification result and comparison with experimental data. Through the same chapter, we have uncovered some of the important points that demand attention during model development. The test on three different learning algorithms led us to choose the suitable one in terms of simplicity, computational time requirement, and accuracy. The test on the estimation of model confidence intervals has been another good consideration. For the case where there is no first principle model, presentation of the performance related models has exposed the capacity of NF based model to the construction of three performance surfaces. The chapter has also demonstrated the dynamic models developed in the frame work of Orthonormal Basis Functions (OBFs). With a latter intent of generating faulty data, the chapter also presented approximate models developed for GTG and HRSG, rsepectively. Summarized form of the outcomes from this chapter are outlined hearunder.

- Semi-empirical model developed for the GTG is validated in Sect. 6.3. Having limited design point data, the proposed model demonstrated fairly accurate result. In fact, it allowed the estimation of the mass flow rate and NOx emission, which are normally not available. The two parameters were predicted setting maximum error tolerance of 0.1% in $\dot{W}_{ele}$ and T_5, respectively.
- One challenge in the design of a FDD system for the HRSG is availability of data for abnormal operating conditions. Aimed at solving the problem, an approximate thermodynamic model has been developed and validated. The validation on T_{g2} and T_{fd3} has demonstrated a relative percentage error of 0.4246 and 1.8776%, respectively. This is under the assumption that 0.1% relative error is used in the

steam flow rate calculation for a given exhaust gas temperature, diverter damper position and steam drum pressure. Given the rough assumptions on approach and pinch temperatures, these much accuracy is quite an outcome.

References

1. Saravanamutto H. I. H., Rogers G. F. C., & Cohen H. (1996). Gas turbine theory (4 ed.). Harlow: Longman Group Limited.
2. Walsh P. P., & Fletcher, P. (2004). Gas turbine performance. New Jersey: Blackwell Science Ltd.
3. Schobeiri, M. (2005). Turbomachinery flow physics and dynamic performance. Berlin: Springer.
4. Razak, A. M. Y. (2007). Industrial gas turbines performance and operability. Cambridge: Woodhead Publishing Limited and CRC Press LLC.
5. Chui, C. K. (1992). An introduction to wavelets. Cambridge: Academic Press Limited.
6. Jackson, J. E. (1991). *A user's guide to principal components* (Vol. 587). USA: Wiley Inc.
7. Oliveira, G. H. C., Campello, R. J. G. B., & Amaral, W. C. (1999). Fuzzy models within orthonormal basis function framework. In *Fuzzy Systems Conference Proceedings, FUZZ-IEEE '99. IEEE International* (Vol. 2, pp. 957–962).
8. Medeiros, A. V., Amaral, W. C., & Campello, R. J. G. B. (2006). GA optimization of OBF TS fuzzy models with linear and non linear local models. In *Ninth Brazilian Symposium on Neural Networks, 2006. SBRN '06* (pp. 66–71).
9. Nelles, O. (1997). Orthonormal basis functions for nonlinear system identification with local linear model trees (LOLIMOT). *IFAC Symposium on System Identification (SYSID)* (pp. 667–672). Fukuoka, Japan: Kitakyushu.
10. Haglind, F. (2010). Variable geometry gas turbines for improving the part-load performance of marine combined cycles - gas turbine performance. *Energy, 31*, 467–476.
11. Coverse, G. L. (1984). Extended parametric representation of compressor fans and turbines. Volume 2: Part user's manual (Parametric Turbine). NASA-CR-174646.
12. Boyce, M. P. (2006). *Gas turbine engineering handbook*. UK: Gulf Professional Publishing.

Chapter 7
Application Studies, Part-II: Fault Detection and Diagnostics

7.1 Introduction

As mentioned at the outset, the main purpose of this research is to develop a Fault Detection and Diagnosis (FDD) method that accommodates nonlinear characteristics of the system, variable operating conditions, signals from all possible sources and a diagnostic result with some imprecision targeted at emulating operators way of describing faulty conditions. To achieve the objective, the whole work was divided into sequence of steps. The first step was about reviewing the signals available for measurement and the faults likely to occur in a CCP. The conclusion from this stage has led us to choose appropriate model identification method. The second step was focused on model identification for normal operating conditions. Thus far, the background theory and review (Chaps. 3 and 4), the proposed FDD techniques (Chap. 5) and the corresponding test and validation result (Chap. 6) have been discussed. In this chapter — which is the result of step four of the research methodology — a test is performed on the application of the proposed FDD system by considering a Cogeneration and District Cooling Plant (CCP) that was elaborated in Chap. 6. Section 7.2 of the chapter concentrates on the use of the methods to the GTG. In the second part (Sect. 7.2), a similar study related to the HRSG and SAC, respectively, will be provided. In presenting the test results, a comparison is made between the suggested approach and the conventional methods like Principal Component Analysis (PCA) and Auto-Associative Neural Network (AANN). The evaluation is carried out in terms of minimum measurement bias that can be detected ($\triangle \delta_{md}$), true detection percentage (τ_{tdp}), true diagnosis percentage (τ_{tip}), and detection delay for an incipient fault. In Sect. 7.3, a summary of the conclusions reached from this chapter is presented.

© Springer International Publishing AG 2018
T.A. Lemma, *A Hybrid Approach for Power Plant Fault Diagnostics*,
Studies in Computational Intelligence 743,
https://doi.org/10.1007/978-3-319-71871-2_7

7.2   Variable Geometry Gas Turbine Generator

As stated in Sect. 1.2, the GTG may experience faults related to system performance, sensors or actuators. In each case the fault may manifest itself as abrupt change or growing with time. The third type of fault, impulsive fault, is also possible. However, it is not that common to dedicate further details. For detecting and further troubleshooting of the responsible part related to plant outage or reduced performance, power plant operators rely on the measured signals. Some of the measurements available for the GTG under consideration are stated in Chap. 2 (see Fig. 2.3). The complete list is available in Appendix F and Table F.6. In the present section, testing of the proposed fault detection and diagnosis method is limited to the same set of sensors.

The whole idea of the FDD system design is to detect the onset of changes in the system as soon as possible, which is before the actual alarm system is initiated. Actual generation of test data is costly and impractical. Hence, the first and main assumption in the following presentation refers to deliberate addition or subtraction of controlled magnitude of sensor errors as percentage of the alarm setting to generate a test data. And latter, the test data is feed to the proposed FDD system to verify that the assumed error can be successfully discerned. Extending this step, different magnitudes of sensor errors will be tested and the minimum that can be detected will be identified. In fact, if there is any abnormal state in the system, it will be first reflected in the measured signals as derangements in one of the above mentioned forms. Actual setpoint data for immediate shutdown of the GTG are given in Appendix F, Table F.10. Latter in the analysis, this information will be used as a reference to set the magnitude of a measurement bias. In Sect. 7.2.1, fault detection test results are presented while in Sect. 7.2.2 the diagnosis tests on the same set of sensors are elaborated.

7.2.1   Fault Detection in the Gas Turbine Generator

The GTG is monitored through sets of signals, some of which related to the working fluid. The rest are for the lube system, generator thermal condition, and bearing vibration monitoring. Those in the gas path are very important to monitor if performance of the system is in close agreement with the expected value. The controller uses measurements of compressor discharge pressure, VIGV position, fuel flow rate, and temperature at the inlet to the 3rd stage of the gas turbine (T_5) to ensure safe running of the GTG according to the required load demand. While this is the main purpose, the same set of signals is important for performance parameter calculations. Besides, the presence of compressor fouling is often verified by a reduced compressor discharge pressure and increase in T_5. Any wrong signal in the control loop has the direct tendency of tripping the plant. Repeated trip causes fatigue failure that may lead to reduced life of the system. In consideration of the facts, it can be said that ensuring the correctness of the measuring sensors is as important as the whole plant itself. In the following sub-section, detection and diagnosis of sensor faults in the gas path system is presented. As abrupt and incipient fault, the following magnitudes of errors are assumed.

For abrupt fault:

$$\Delta x(p) = \begin{cases} 0, & for\ p < 300 \\ \Delta x_{max}, & for\ p > 300 \end{cases} \tag{7.1}$$

For incipient fault:

$$\Delta x(p) = \begin{cases} \frac{\Delta x_{max}}{200}(p - 200), & for\ 200 \leq p \leq 400 \\ \Delta x_{max}, & for\ p > 400 \end{cases} \tag{7.2}$$

where Δx_{max} is the maximum change assumed for a particular sensor and p is the data point.

Gas Path Related Sensor Faults

In order to demonstrate the use of the FDD system to GTG gas path sensors, a total of 29 test cases are considered. The first test is for fault free condition while the rest are corresponding to each sensor fault (refer to Tables H.1 and H.2). In the sequel, the sensor for pressure drop in the compressor suction duct is selected for detail discussion. Basically, this sensor is installed to monitor state of the air filter in the suction duct. The pressure drop increases when it is dust clogged. A value reaching 1.9 kPa differential pressure causes automatic plant shutdown. For the PCA and NF approach, the models developed and validated in Chap. 6 are used. The relations for the calculation of PCA model confidence interval can be seen in Appendix C. Details of the AANN model structure are elaborated in the same appendix. AANN models are arranged for the gas path, lube system, generator coils, and turbine inlet temperatures, respectively. In Table 7.1 are given detail of the model structure in the AANN based models. For each case, Levenberg Marquardt (LM) algorithm is employed as training method. The detection test result for fault free condition is indicated in Figs. 7.1, 7.2, and 7.3. As shown in the figures, all the three modelling techniques correctly represented the normal state. While the NF approach full contained the prediction, the PCA and AANN models relatively show false alarms. This, however, can be improved by using relaxed confidence levels.

Table 7.1 Number of nodes in the AANN models for GTG

Model	AANN structure					Learning algorithm
	nx	nl	nc	nm	ny	
Gas path	13	8	2	8	13	LM
Lube system	6	8	2	8	6	LM
Generator coils	3	8	2	8	3	LM
T_5 sensors	7	8	2	8	7	LM

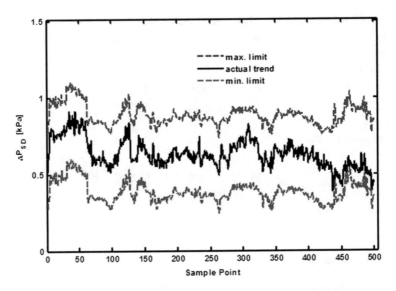

Fig. 7.1 NF based fault detection result for normal operating condition: $\triangle$PSD

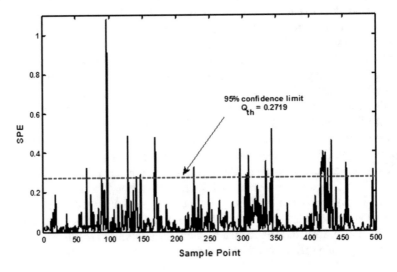

Fig. 7.2 PCA based fault detection result for normal operating condition: $\triangle$PSD

In the second test, 20% of the alarm setting is added to the data points from 200 onwards to create a bias error. As demonstrated in Figs. 7.4, 7.5 and 7.6, the calculated residuals crossed the model thresholds in all the models except the result from PCA. This shows how competent the NF approach is with respect to PCA.

In the following, instead of considering the whole sensors involved in the gas path, we would rather focus on T_1, P_2, P_{vc} and T_{enc}. T_1 and T_{enc} are outside the control

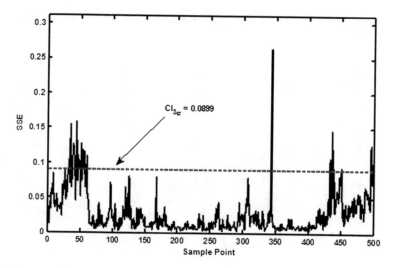

Fig. 7.3 AANN based fault detection result for normal operating condition: △PSD

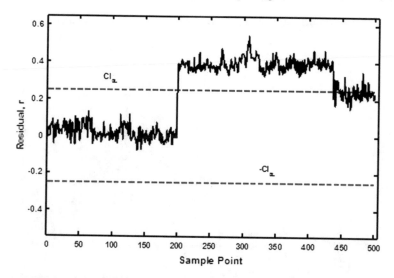

Fig. 7.4 NF based fault detection: 20% bias in △PSD

loop. While a false reading for T_1 may cause wrong energy audit calculations, T_{eco} is limited to the safety inside the room that hosts the GTG. In some designs P_2 is considered as part of the main control loop. Here, since we had limited information to confirm that, we have assumed P_2 as having minimal contribution to the feedbacks to the control system. In the sequel, detection of abrupt and incipient faults, respectively, corresponding to each of the four signals is studied.

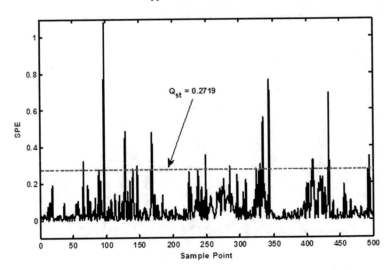

Fig. 7.5 PCA based fault detection: 20% bias in △PSD

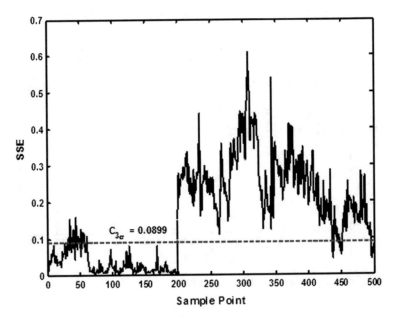

Fig. 7.6 AANN based fault detection: 20% bias in △PSD

Using PCA, a 10% change in compressor inlet temperature can be detected with detection ratio of 98.33%. But, the diagnosis is difficult to decide for the Q-static plot indicates two possible suspects: T_1 and T_{enc}. The minimum change that can be detected by PCA model is about 6.5%. For AANN approach to work, the change has to be bigger than 19.5%, which is relatively high. On the other hand, the NF approach managed to detect sensor bias as small as 9.2% while demonstrating 61.67% true detection ratio for an assumed fault of 10% with respect to the alarm setting.

In the compressor discharge pressure, the three approaches performed well for the case where the change is about 10% of the alarm setting. However, the NF approach is more efficient as it is capable of detecting changes as small as 3.1%. In the case of fuel pressure (P_{vc}) sensor fault, the NF approach worked efficiently though it is next to PCA.

Regarding the turbine room temperature (T_{enc}), after grouping it in the gas path model, the PCA method is capable of detecting a 10% change. If the AANN method is used, the bias has to be greater than 12.5%, which is relatively higher. In the NF approach, the model is defined as a function of VIGV position and fuel flow rate, Appendix H and Table H.3. The true detection percentage is less than 50% if the change in the sensor is lower than 10%. However, it performed better than AANN.

Summary of the test results for the four sensors are given in Tables 7.2 and 7.3. The second table outlines the detection delay for an incipient fault governed by (7.2). In all the tests, NF models outperformed the models based on AANN.

Detection of Changes in the Turbine Stage Temperatures

Whenever there is a problem in the combustion chamber, the situation reveals itself as changes in the temperature measured at the inlet to the 3rd stage of the gas turbine. Excessive temperature could lead to system trip. This is so in all operating regions. Basically, the same temperature averaged over the six sensors arranger at equal circumferential angles and corrected for compressor inlet temperature is one of the inputs to the controller sought to successfully run the GTG in the SoLoNox

Table 7.2 Minimum detectable abrupt faults and percentage detection for 10% change: gas path sensors

Sensors	Method								
	PCA			AANN			NF		
	$\Delta\delta_{md}$	τ_{tdp}	τ_{tip}	$\Delta\delta_{md}$	τ_{tdp}	τ_{tip}	$\Delta\delta_{md}$	τ_{tdp}	τ_{tip}
T_1	6.5	98.33	–	19.5	4	–	9.2	61.67	100
P_2	4.1	100	100	13.5	10	100	3.1	100	99.5
P_{vc}	0.29	100	100	1.1	100	100	0.89	100	100
T_{enc}	3.8	100	100	12.5	12	–	11	34.50	–

Table 7.3 Detection delay for an assumed incipient fault: gas path sensors

Sensors	Detection delay		
	PCA	AANN	NF
T_1	44	244	131
P_2	42	135	57
P_{vc}	6	26	20
T_{enc}	33	143	137

mode. This section addresses the use of the FDD system to the health monitoring of T_5 sensors. Analogous to the previous tests, models are first developed applying NF, PCA and AANN methods, respectively. Then, the models are applied to the faulty data that is created by implanting abrupt and incipient changes to the normal operation data. While setting a maximum change of 10%, we used (7.1) and (7.2) to generate the required test data.

In Table 7.4 are outlined the test results the minimum that can be detected ($\Delta\delta_{md}$), true detection percentage (τ_{tdp}), and true diagnosis percentage (τ_{tip}) corresponding to an abrupt fault. FDD based on NF approach exhibited minimum detection percentage relative to AANN method. Though PCA is more sensitive to small changes, the true diagnosis percentage deteriorated with bias magnitude. In the NF method, however, the high diagnosis ratio still persists even though the change is near to 2%. Regarding the detection delay, the NF is as good as PCA, Table 7.5. In the NF method the input-output relations are identified according to the relations depicted in Appendix H, Tables H.3 and H.4.

Lube System Sensor Fault in the SoLoNOx Region

This section is included to test if the proposed fault detection method is equally applicable to the sensors in the lube system. The lube system is one of the auxiliary systems vital for safe running of the GTG (see Sect. 6.2.1). Apart from lube oil

Table 7.4 Minimum detectable abrupt faults and percentage detection: gas turbine 3rd stage inlet temperature sensors (T_5)

Sensors	Method								
	PCA			AANN			NF		
	$\Delta\delta_{md}$	τ_{tdp}	τ_{tip}	$\Delta\delta_{md}$	τ_{tdp}	τ_{tip}	$\Delta\delta_{md}$	τ_{tdp}	τ_{tip}
T_{c1}	2.5	100	–	7.8	92.67	100	3.7	100	100
T_{c2}	1.5	100	–	8.7	71	100	3.3	100	98.33
T_{c3}	3	100	–	9.6	62	100	2.6	100	100
T_{c4}	2.5	93.67	–	7.7	94	100	1.9	100	100
T_{c5}	1.9	100	–	7.7	98.67	100	2.15	100	100

Table 7.5 Minimum detectable abrupt faults and percentage detection: gas turbine 3rd stage inlet temperature sensors

Sensors	Detection delay (min)		
	PCA	AANN	NF
T_{c1}	27	95	51
T_{c2}	18	108	40
T_{c3}	24	139	35
T_{c4}	24	95	26
T_{c5}	18	109	40
T_{c6}	24	95	64

monitoring for wear rate prediction, temperatures and pressure data at different locations are also monitored to ensure that the system is safe to run. High lube oil temperature is a good indicator of excessive vibration and even oil leak. A drop in oil pressure also signals oil leak. Valve and lube oil pressure controller or actuator malfunctions are often detected by the change in the trend in the pressure and temperature signals. While a fault in a gas path sensor outside the control loop is some how less destructive and tolerable, the consequence from a missed fault in the lube system is quit destructive; it may lead to catastrophic failures with expensive maintenance cost. Related to the lube system, there are five temperature sensors and one pressure sensor (see Appendix F and Table F.6).

In Sect. 6.3.1, we have presented the validation result on the PCA model developed for the lube system. Two principal components were enough to capture 99.7% of the data variance for normal operating conditions. Structure of the AANN model for the same system is given in Table 7.1. To use the proposed method, NF models are also developed choosing the inputs for each model in such a way that the model structure ought to be suitable for fault diagnosis. Detail of the input-output relations can be seen from Appendix H and Table H.3. The number of rules for each sensor is given in Table 7.6.

What we did in the FDD test is that, we used the same assumption as stated by (7.1), to determine the minimum sensor bias that can be detected, the true detection ratio and the true diagnosis percentage. In the incipient fault detection, which is the second part of the test, the detection delay behind each sensor is estimated. Summary of the test results can be found in Tables 7.7 and 7.8. As can be seen, in all the test cases, the NF based approach demonstrated a performance better than PCA and AANN methods. Except for two sensors - T_{lb3} and P_{lb1} - the diagnosis percentage is also higher than its counterpart. In Fig. 7.7 are shown plots of the residuals corresponding to temperature sensor T_{lb1}. In the faulty region, three of the residuals r_1, r_2 and r_6 have crossed their respective model threshold for normal operating condition. Based on this observation, it can be concluded that the NF based method is efficient enough to be used for monitoring the state of temperature and pressure sensors in the lube system.

Table 7.6 Training and test performance for sensor in the lube system: NF models for normal operating condition

Parameter	Number of rules	Training data			Test data		
		RMSE	AIC	VAF	RMSE	AIC	VAF
T_{lb1}	6	0.0046	−10.71	98.6	0.0045	−10.78	98.38
T_{lb2}	3	0.0042	−10.91	98.15	0.043	−10.86	97.67
T_{lb3}	5	0.0095	−9.28	93.1	0.0103	−9.12	90.48
T_{lb4}	3	0.0024	−12.0	98.45	0.0024	−12.1	98.32
T_{lb5}	3	0.0026	−11.89	98.47	0.0026	−11.91	98.29
P_{lb1}	3	0.0542	−5.81	−55.57	0.0662	−5.4	36

Table 7.7 Minimum detectable abrupt faults and percentage detection: lube system sensors

Sensors	Model identification method								
	PCA			AANN			NF		
	$\triangle\delta_{md}$	τ_{tdp}	τ_{tip}	$\triangle\delta_{md}$	τ_{tdp}	τ_{tip}	$\triangle\delta_{md}$	τ_{tdp}	τ_{tip}
T_{lb1}	2.6	100	100	2.6	100		1.7	100	100
T_{lb2}	3.0	100	100	2.8	100		1.5	100	100
T_{lb3}	2.4	100	100	2.1	100		2.8	100	–
T_{lb4}	1.0	100	100	0.85	100		0.52	100	100
T_{lb5}	1.0	100	100	1.05	100		0.51	100	100
P_{lb1}	5.5	100	0	4.5	100		6.6	100	–

Table 7.8 Detection delay for an assumed incipient fault: lube system sensors

Sensors	Detection delay, [Number of samples]		
	PCA	AANN	NF
T_{lb1}	56	48	24
T_{lb2}	34	51	21
T_{lb3}	27	39	41
T_{lb4}	16	22	22
T_{lb5}	19	18	22
P_{lb1}	57	89	33

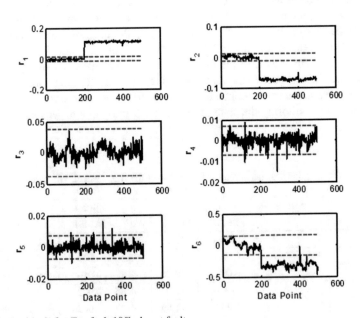

Fig. 7.7 Residuals for T_{lb1} fault 10% abrupt fault

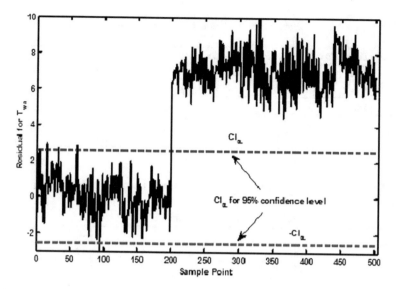

Fig. 7.8 Residual and confidence limits for T_{wa} sensor: 10% abrupt fault

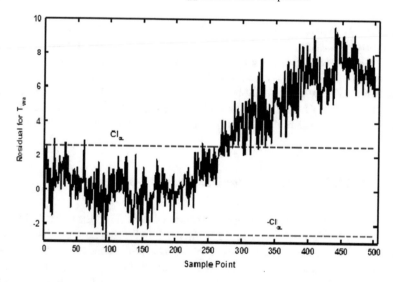

Fig. 7.9 Residual and CI for T_{wa} sensor: maximum 10% incipient fault

Generator Winding Coils

The subject of this section is the detection of abrupt and incipient faults in the generator winding coil temperature sensors. The fault in the sensors follows the assumptions made in (7.1) and (7.2). Figures 7.8 and 7.9 outline the trends related to temperature sensor T_{wa} and corresponding to abrupt and incipient faults, respectively. The Figures also show the upper and lower model thresholds for 95% confidence

Table 7.9 Minimum detectable abrupt faults and percentage detection: generator coil temperature sensors

Sensors	Detection delay, [Number of samples]								
	PCA			AANN			NF		
	$\Delta\delta_{md}$	τ_{tdp}	τ_{tip}	$\Delta\delta_{md}$	τ_{tdp}	τ_{tip}	$\Delta\delta_{md}$	τ_{tdp}	τ_{tip}
T_{wa}	3.2	100	100	4.6	100	100	3.7	100	100
T_{wb}	2.7	100	100	4.2	100	100	3.7	100	96.67
T_{wc}	2.8	100	100	4.5	100	100	2.4	100	99

Table 7.10 Detection delay for an assumed incipient fault: lube system and generator coil sensors

Sensors	Detection delay (min)		
	PCA	AANN	NF
T_{wa}	40	69	40
T_{wb}	20	61	49
T_{wc}	40	64	28

level. If NF method is applied to monitor T_{wa}, a change as small as 3.7% is detectable with a corresponding τ_{tdp} of 54.5%. The same fault is discernable if binary evaluation of the residual signals is applied. When AANN is used, the minimum that can be detected contracts to 4.6%. A similar behavior is seen in T_{wa} and T_{wc}. A summary of the test results are given in Tables 7.9 and 7.10. In terms of detection delay, NF is better than AANN. Overall, it can be said that the proposed NF approach is adequate enough to detect faults in the generator coil temperature sensors.

7.2.2 Fault Diagnosis in the Gas Turbine Generator

In this section, we show application of the proposed fault diagnosis system. Details of the method can be seen from Chap. 4 and Sect. 4.4. In the first part, the diagnosis test focuses on the lube system. Secondly, a similar analysis is conducted on the sensors involved in the generator coils. The reason behind choosing the auxiliary systems over the gas path sensors is due to the fact that the latter is featured by thirteen sensors and plotting the residuals, binary signals and fuzzy fault activation levels may make the presentation awkward. Since the FDD method is the same regardless any of the subsystems, we argue that this decision doesn't create any problems in exposing the anticipated characteristics.

Fault diagnosis is a two step process. In the first step, the calculated residuals are evaluated by a pre designed residual evaluator. In our case, two options are available: binary evaluation and fuzzy evaluation (see Sect. 4.4). In the second step, the signals from the first step are used in a Binary Diagnostic Matrix (BDM). The

BDM designed for the GTG is depicted in Appendix H and Fig. H.1. Since the BDM carries the signature for the possible faults in the GTG, the execution of step two results in a specific signature corresponding to the likely fault. In case of using the fuzzy evaluation, support of the fuzzy functions (see Sect. 5.2.3) are assumed as a $=$ 1.5 and b $= 0.5$. This according to our design remains valid irrespective of the type of sensor.

Fault Diagnosis in the Lube System

As discussed in Sect. 7.2.1, the lube system is featured by six sensors. For each sensor, the proposed fault detection system has shown interesting results. What is presented hereunder is the extension to fault diagnosis. Referring to the BDM given in Appendix H, the set of if-then rules that is designed to serve the purpose of fault diagnosis in terms of binary signals in the lube system is as follows:

$$If\ (s_1 = 0) \cap (s_2 = 0) \cap (s_3 = 0) \cap (s_4 = 0) \cap (s_5 = 0) \cap (s_6 = 0)\ then\ \mathbf{f_0}$$

$$If\ (s_1 = 0) \cap (s_2 = 0) \cap (s_3 = 0) \cap (s_4 = 0) \cap (s_5 = 0) \cap (s_6 = 0)\ then\ \mathbf{f_1}$$

$$If\ (s_1 = 0) \cap (s_2 = 0) \cap (s_3 = 0) \cap (s_4 = 0) \cap (s_5 = 0) \cap (s_6 = 0)\ then\ \mathbf{f_2}$$

$$If\ (s_1 = 0) \cap (s_2 = 0) \cap (s_3 = 0) \cap (s_4 = 0) \cap (s_5 = 0) \cap (s_6 = 0)\ then\ \mathbf{f_3}$$

$$If\ (s_1 = 0) \cap (s_2 = 0) \cap (s_3 = 0) \cap (s_4 = 0) \cap (s_5 = 0) \cap (s_6 = 0)\ then\ \mathbf{f_4}$$

$$If\ (s_1 = 0) \cap (s_2 = 0) \cap (s_3 = 0) \cap (s_4 = 0) \cap (s_5 = 0) \cap (s_6 = 0)\ then\ \mathbf{f_5}$$

$$If\ (s_1 = 0) \cap (s_2 = 0) \cap (s_3 = 0) \cap (s_4 = 0) \cap (s_5 = 0) \cap (s_6 = 0)\ then\ \mathbf{f_6}$$

The corresponding fuzzy rules can also be constructed. Adopting the design discussed in Sects. 4.4.2 and 5.2.3, we have

$$If\ (s_1 = \mu_{NZ.1}) \otimes (s_2 = \mu_{NZ.2}) \otimes (s_3 = \mu_{NZ.3}) \otimes \dots$$
$$(s_4 = \mu_{NZ.4}) \otimes (s_5 = \mu_{NZ.5}) \otimes (s_6 = \mu_{NZ.6})\ then\ \mathbf{f_0}$$

$$If\ (s_1 = \mu_{Z.1}) \otimes (s_2 = \mu_{Z.2}) \otimes (s_3 = \mu_{NZ.3}) \otimes \dots$$
$$(s_4 = \mu_{NZ.4}) \otimes (s_5 = \mu_{NZ.5}) \otimes (s_6 = \mu_{NZ.6})\ then\ \mathbf{f_1}$$

$$If\ (s_1 = \mu_{Z.1}) \otimes (s_2 = \mu_{Z.2}) \otimes (s_3 = \mu_{NZ.3}) \otimes \dots$$
$$(s_4 = \mu_{NZ.4}) \otimes (s_5 = \mu_{NZ.5}) \otimes (s_6 = \mu_{NZ.6})\ then\ \mathbf{f_2}$$

$$\textit{If } (s_1 = \mu_{NZ.1}) \otimes \ (s_2 = \mu_{Z.2}) \otimes (s_3 = \mu_{Z.3}) \otimes \ldots$$
$$(s_4 = \mu_{Z.4}) \otimes (s_5 = \mu_{NZ.5}) \otimes (s_6 = \mu_{NZ.6}) \textit{ then } \mathbf{f}_3$$

$$\textit{If } (s_1 = \mu_{NZ.1}) \otimes \ (s_2 = \mu_{NZ.2}) \otimes (s_3 = \mu_{Z.3}) \otimes \ldots$$
$$(s_4 = \mu_{Z.4}) \otimes (s_5 = \mu_{Z.5}) \otimes (s_6 = \mu_{NZ.6}) \textit{ then } \mathbf{f}_4$$

$$\textit{If } (s_1 = \mu_{NZ.1}) \otimes \ (s_2 = \mu_{NZ.2}) \otimes (s_3 = \mu_{NZ.3}) \otimes \ldots$$
$$(s_4 = \mu_{Z.4}) \otimes (s_5 = \mu_{Z.5}) \otimes (s_6 = \mu_{Z.6}) \textit{ then } \mathbf{f}_5$$

$$\textit{If } (s_1 = \mu_{Z.1}) \otimes \ (s_2 = \mu_{NZ.2}) \otimes (s_3 = \mu_{NZ.3}) \otimes \ldots$$
$$(s_4 = \mu_{NZ.4}) \otimes (s_5 = \mu_{Z.5}) \otimes (s_6 = \mu_{Z.6}) \textit{ then } \mathbf{f}_6$$

Now, we apply the rules to the residual signals for T_{lb1} (see Fig. 7.7). The binary evaluation led to the plots as indicated in Fig. 7.10 while the fuzzy evaluation revealed the most likely fault as indicated in Fig. 7.11. According to the rules for fault diagnosis system, the fault is linked to T_{lb1} sensor if the signature is given by:

$$\mathbf{V} = [1 \quad 1 \quad 0 \quad 0 \quad 0 \quad 1]^T \tag{7.3}$$

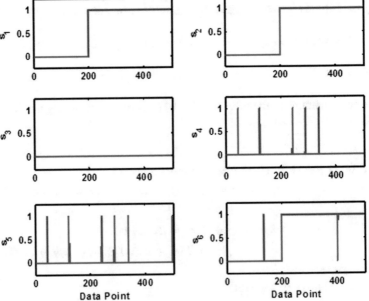

Fig. 7.10 Binary fault signals for T_{lb1} fault

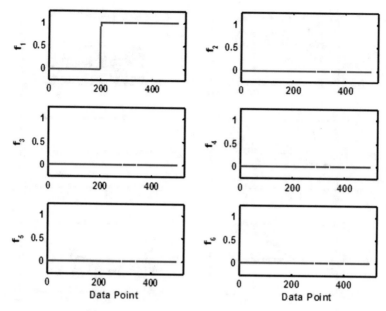

Fig. 7.11 Fuzzy fault activation level for T_{lb1} fault

Referring to Fig. 7.10, for data samples from 200 and on wards, the trends indeed reflect the signature given by (7.3). Therefore, it can be concluded that the proposed diagnosis system has efficiently diagnosed the fault. The fuzz method also shows that, Fig. 7.11, the most likely fault is f_1, which stands for a fault in T_{lb1}.

As a second test, we considered fault diagnosis in the lube oil temperature sensor at the generator exciter bearing. This temperature is designated by T_{lb4} (see Appendix G). Instead of considering abrupt fault, the test is done for an incipient fault with the bias variation governed by (7.2). Assuming the maximum added change is 10%, the calculated residuals for all the sensors in the lube system feature as shown in Fig. 7.12. Out of six calculated residuals three of them crossed the adaptive threshold estimated to define normal operating conditions. Binary evaluation of the residuals reveals that the fault signature for the data samples starting at about 250 and beyond is $V = [1 \ 1 \ 0 \ 0 \ 0 \ 1]^T$ (see Fig. 7.13), which according to the aforementioned rules reference to fault type f_4. Basically, the conclusion is supported by the diagnosis result from fuzzy evaluation of the residual signals, Fig. 7.14. In consideration of the result from analysis of abrupt and incipient faults in the two sensors, we conclude that the proposed diagnosis system is capable of identifying a sensor fault in the lube system.

Generator Winding Coil Temperature Fault in the SoLoNOx Operating Region

The diagnosis in the generator coils is performed based on the BDM given in Appendix H and Fig. H.1. In Figs. 7.15, 7.16 and 7.17 are depicted trends of residuals, binary signals and fuzzy activation level, respectively, corresponding to a fault in

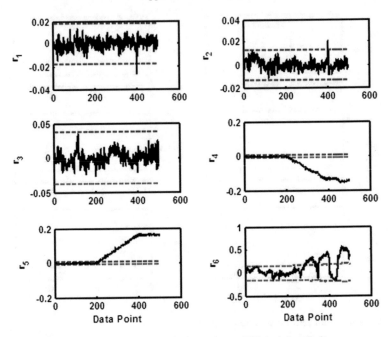

Fig. 7.12 Residuals corresponding to T_{lb4} fault maximum 10% incipient fault

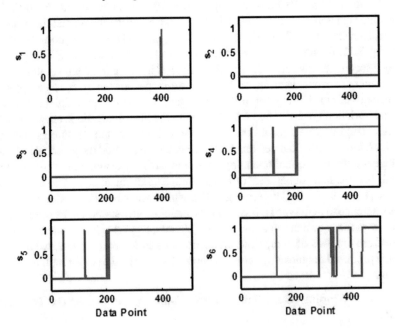

Fig. 7.13 Binary fault signals for T_{lb4} fault

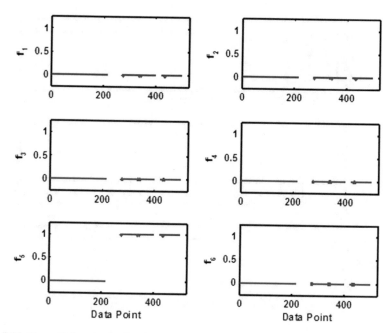

Fig. 7.14   Fuzzy fault activation level for T_{lb4} fault

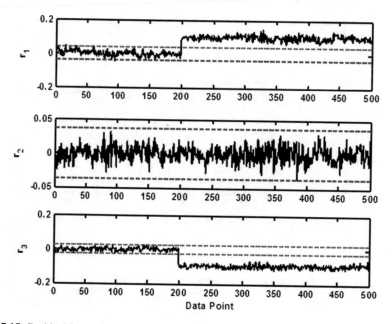

Fig. 7.15   Residual for an abrupt fault in T_{wa} sensor

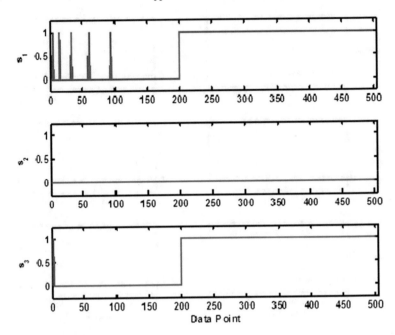

Fig. 7.16 Binary signals for an abrupt fault in T_{wa} sensor

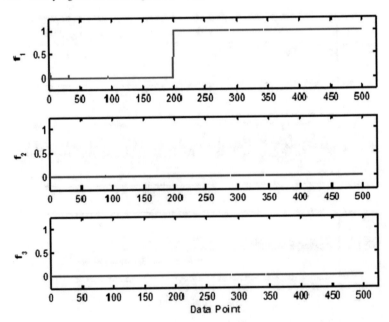

Fig. 7.17 Fuzzy activation Level for an abrupt fault in T_{wa} sensor

temperature sensor T_{wa}. According to the BDM, a fault signature of $[1 \quad 0 \quad 1]^T$ signifies a fault in T_{wa}. This is shown in Fig. 7.16 for data samples 200 to the 500. In Fig. 7.17, the fault fuzzy activation for T_{wa} is higher than the rest, which demonstrated that a fault in T_{wa} is the most likely fault. From the observations, it can be said that the fault in Twa is well discerned. We did a similar test on the other two sensors and the outcomes were as anticipated. We conclude that the proposed FDD system is also applicable for monitoring the condition of gas turbine generator coils.

7.3 12 Ton/Hr Rated Heat Recovery Steam Generator

The common faults in the HRSG are sensor faults, actuator faults and performance faults: fouling in the evaporator, economizer and blow-down heat exchanger. Delay in the opening of the diverter damper due to actuator fault could cause GTG trip. Fouling in the HRSG, apart from affecting the performance of the HRSG itself, may lead to increase in GTG back pressure, which has a negative effect on the performance of the GTG. In Sect. 7.3.1, the detection of different magnitudes of abrupt and impending changes in the system are addressed. In Sect. 7.3.2, the diagnosis result applying the NF method as developed in Chaps. 3–5 is outlined.

7.3.1 Fault Detection in the HRSG

The detection of faults in the HRSG follows the procedure developed in Chap. 5 and Sect. 5.2.2. The model develop for normal operation are used to predict the outputs for a new set of input, which is the same input experienced by the system. Residuals are, then, calculated between the prediction results and actual data. Depending up on the size of the residual with respect to the threshold for normal operating condition, the trend is classified as either normal or faulty. Unlike the steps in the GTG, the approach followed in this section is to use NF approach alone. For normal operating condition, measurement errors (f_u and f_y) are assumed zero.

In total, twelve measurement sensors are available for the HRSG. Each sensor provides a signal on the state of itself and a propagated fault. A propagated fault is when there is a process fault and a sensor is involved in the process signalling its presence. The list of sensor faults and the corresponding fault detection tests are listed in Appendix H (see Tables H.5 and H.6).

To test the fault detection and diagnosis system, pre-determined bias errors are added to the measurements corresponding to normal operating conditions. In fact the procedure we are following is similar to that used for GTG. The maximum value of the bias errors assumed for the measurements are 10% of the design value stated in Appendix F and Table F.3. In testing the FDD scheme, only one sensor at a time is biased as multiple sensor faults are not frequent. Besides, only sensors not involved in the feedback control loop are biased for it is impossible to simulate the real operation of the HRSG by simply adding a bias error to a measurement used in a feedback control loop without affecting the other measurements. The sensors considered for the test are: hot gas inlet temperature (T_{g1}), gas temperature at the inlet to the economizer (T_{g2}), feed water inlet temperature to the economizer (T_{fd2}), temperature of feed water leaving the economizer (T_{fd3}), and temperature of exhaust gas in the by-pass stack (T_{g4}). In total, there are six test cases. Out of six, one is for normal operating condition.

Two kinds of faults are tested. The first is an abrupt fault with a maximum bias of 10% with respect to the design value. To create the test data, a fixed bias is added to the normal operation data that falls in the range of 100 to 400 sample points, with the total sample being 400. The decision on the location of added bias is arbitrary for the whole purpose is to test the performance of the proposed FDD system.

Regarding the incipient fault, the maximum assumed derangement is still 10% with respect to the design value. However, the gradual increment in the bias is adjusted to span data points from 100 to 300, with the maximum bias being added to the rest of the data. The assumed governing equation for the bias variation in the incipient fault region is given by:

$$\Delta x(p) = \begin{cases} \frac{\Delta x_{max}}{200}(p - 100), \ for \ 100 \leq p \leq 300 \\ \Delta x_{max}, for \ p > 300 \end{cases} \tag{7.4}$$

where, Δx_{max} is the maximum assumed bias for a particular sensor and p is the data point.

Plots of the residuals calculated for the normal operating condition are shown in Fig. 7.18. As indicated in the Figure, the residuals are all less than their respective model confidence limits, which are estimated for 95% confidence level. Hence, it can be said that there is no fault in the system.

The test results with the assumed bias errors and for the five test cases are given in Table 7.11. In the $\Delta x \geq 10\%$ region, the true detection percentage (τ_{tdp}) and true diagnosis percentage (τ_{tip}) for each case is 100%. Figure 7.19 provides the residual plots for the test data generated by adding a bias error of 5.4 °C to T_{g1} normal operation data. In the faulty region, the residual for T_{g1} is above the model confidence limit. This demonstrates that the fault is detected correctly. A similar study is performed for the other sensors as well. For T_{g1} a minimum bias error of 5.5% can be detected while the diagnosis result is much discernable if a bias error of higher than 7.5% is considered. In Table 7.11, the numbers in the curled bracket indicate the true detection percentage and true diagnosis ratio corresponding to minimum detection percentage. In the tests for incipient fault, the detection delay is a function of the minimum bias that can be detected. As can be seen, for lower detection percentage, the detection delay is lower. Based on the analysis results, it can be said that the proposed FDD is equally applicable to sensor fault detection in the HRSG.

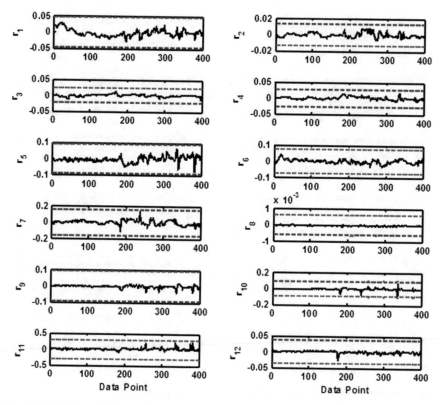

Fig. 7.18 HRSG residual plots for normal operating condition no fault

Table 7.11 Minimum detectable abrupt faults, percentage detection, percentage diagnosis and detection delay for HRSG sensors: NF based approach

Sensors	FDD performance parameter			
	τ_{md}	τ_{tdp}	τ_{tip}	τ_{dd}
T_{g1}	5.5	100(65.7)	100(65.7)	89
T_{g2}	1.4	100(54.0)	100(54.0)	32
T_{fd2}	2.2	100(59.7)	100(0)	40
T_{fd3}	2.8	100(58.3)	100(58.3)	57
T_{g4}	7.6	97.7(53.7)	97.7(53.7)	145

7.3.2   Fault Diagnosis in the HRSG

The first step in fault diagnosis is designing the Binary Diagnostic Matrix (BDM). Once the BDM is formulated, the corresponding Fuzzy Diagnostic Matrix (FDM) can be arranged assuming suitable fuzzy membership functions. For the HRSG, since

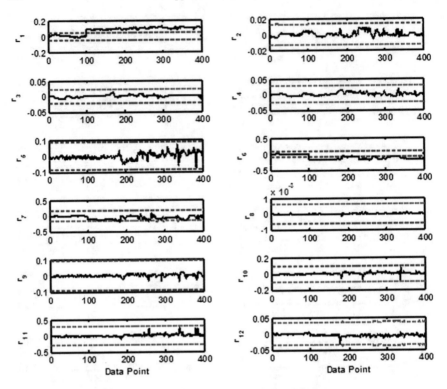

Fig. 7.19 HRSG residual plots for T_{g1} sensor fault -10% abrupt fault

we have twelve measured variables, the size of the BMD is at least 12 rows by 13 columns. The rows indicate the number of sensors while the columns are for the type of fault and its signature. Considering sensor faults only, the designed BDM for the HRSG is as demonstrated in Fig. 7.20. The size of the matrices increases if process faults are included. The FDM constructed on the bases of the BDM is documented in Appendix H and Fig. H.2. In the following the diagnosis of T_{g2}, T_{fd2}, T_{fd3} and T_{g4} sensors applying the designed BDM and FDM are presented. But, first, lets see the fuzzy rules behind the faults included in the BDM or FDM structure. Note that the rules are constructed on the bases of the BDM and the fuzzy functions elaborated in Chap. 4 and Sect. 4.4.2.

For no fault, the rule resembles:

$$If \ (s_1 = \mu_{Z,1}) \otimes (s_= \mu_{Z,2}) \otimes \ (s_3 = \mu_{Z,3}) \otimes (s_4 = \mu_{Z,4}) \otimes (s_5 = \mu_{Z,5})$$
$$\otimes (s_6 = \mu_{Z,6}) \otimes (s_7 = \mu_{Z,7}) \ \otimes (s_8 = \mu_{Z,8}) \otimes (s_9 = \mu_{Z,9}) \otimes (s_10 = \mu_{Z,10})$$
$$\otimes (s_11 = \mu_{Z,11}) \otimes (s_12 = \mu_{Z,12}) \ then \ \mathbf{f}_0$$

Fig. 7.20 Binary diagnostic matrix for the HRSG

R	f_0	f_1	f_2	f_3	f_4	f_5	f_6	f_7	f_8	f_9	f_{10}	f_{11}	f_{12}
S_1	1			1	1		1						
S_2	1	1		1			1						
S_3					1		1		1				
S_4					1	1	1						
S_5							1					1	
S_6		1					1						
S_7								1			1		1
S_8								1	1				
S_9								1		1			
S_{10}							1			1	1		
S_{11}							1					1	
S_{12}								1	1				1

For the rest of the sensors, we have: Rule-2:

$$If\ (s_1 = \mu_{NZ,1}) \otimes (s_= \mu_{Z,2}) \otimes (s_3 = \mu_{Z,3}) \otimes (s_4 = \mu_{Z,4}) \otimes (s_5 = \mu_{Z,5})$$
$$\otimes (s_6 = \mu_{NZ,6}) \otimes (s_7 = \mu_{Z,7}) \otimes (s_8 = \mu_{Z,8}) \otimes (s_9 = \mu_{Z,9}) \otimes (s_10 = \mu_{Z,10})$$
$$\otimes (s_11 = \mu_{Z,11}) \otimes (s_12 = \mu_{Z,12})\ then\ \mathbf{f}_1$$

Rule-3:

$$If\ (s_1 = \mu_{Z,1}) \otimes (s_= \mu_{NZ,2}) \otimes (s_3 = \mu_{Z,3}) \otimes (s_4 = \mu_{Z,4}) \otimes (s_5 = \mu_{Z,5})$$
$$\otimes (s_6 = \mu_{Z,6}) \otimes (s_7 = \mu_{Z,7}) \otimes (s_8 = \mu_{Z,8}) \otimes (s_9 = \mu_{Z,9}) \otimes (s_10 = \mu_{Z,10})$$
$$\otimes (s_11 = \mu_{Z,11}) \otimes (s_12 = \mu_{Z,12})\ then\ \mathbf{f}_2$$

Rule-4:

$$If\ (s_1 = \mu_{Z,1}) \otimes (s_= \mu_{Z,2}) \otimes (s_3 = \mu_{NZ,3}) \otimes (s_4 = \mu_{NZ,4}) \otimes (s_5 = \mu_{Z,5})$$
$$\otimes (s_6 = \mu_{Z,6}) \otimes (s_7 = \mu_{Z,7}) \otimes (s_8 = \mu_{Z,8}) \otimes (s_9 = \mu_{Z,9}) \otimes (s_10 = \mu_{Z,10})$$
$$\otimes (s_11 = \mu_{Z,11}) \otimes (s_12 = \mu_{Z,12})\ then\ \mathbf{f}_3$$

Rule-5:

$$If\ (s_1 = \mu_{NZ,1}) \otimes (s_= \mu_{NZ,2}) \otimes (s_3 = \mu_{Z,3}) \otimes (s_4 = \mu_{NZ,4}) \otimes (s_5 = \mu_{Z,5})$$
$$\otimes (s_6 = \mu_{Z,6}) \otimes (s_7 = \mu_{Z,7}) \otimes (s_8 = \mu_{Z,8}) \otimes (s_9 = \mu_{Z,9}) \otimes (s_10 = \mu_{Z,10})$$
$$\otimes (s_11 = \mu_{Z,11}) \otimes (s_12 = \mu_{Z,12})\ then\ \mathbf{f}_4$$

Rule-6:

$$If\ (s_1 = \mu_{NZ,1}) \otimes (s_{=}\mu_{Z,2}) \otimes (s_3 = \mu_{NZ,3}) \otimes (s_4 = \mu_{NZ,4}) \otimes (s_5 = \mu_{NZ,5})$$
$$\otimes(s_6 = \mu_{Z,6}) \otimes (s_7 = \mu_{Z,7}) \otimes (s_8 = \mu_{Z,8}) \otimes (s_9 = \mu_{Z,9}) \otimes (s_10 = \mu_{NZ,10})$$
$$\otimes(s_11 = \mu_{Z,11}) \otimes (s_12 = \mu_{Z,12})\ then\ \mathbf{f}_5$$

Rule-7:

$$If\ (s_1 = \mu_{Z,1}) \otimes (s_{=}\mu_{Z,2}) \otimes (s_3 = \mu_{Z,3}) \otimes (s_4 = \mu_{Z,4}) \otimes (s_5 = \mu_{Z,5})$$
$$\otimes(s_6 = \mu_{NZ,6}) \otimes (s_7 = \mu_{Z,7}) \otimes (s_8 = \mu_{Z,8}) \otimes (s_9 = \mu_{Z,9}) \otimes (s_10 = \mu_{Z,10})$$
$$\otimes(s_11 = \mu_{Z,11}) \otimes (s_12 = \mu_{Z,12})\ then\ \mathbf{f}_6$$

Rule-8:

$$If\ (s_1 = \mu_{NZ,1}) \otimes (s_{=}\mu_{NZ,2}) \otimes (s_3 = \mu_{Z,3}) \otimes (s_4 = \mu_{Z,4}) \otimes (s_5 = \mu_{Z,5})$$
$$\otimes(s_6 = \mu_{NZ,6}) \otimes (s_7 = \mu_{NZ,7}) \otimes (s_8 = \mu_{NZ,8}) \otimes (s_9 = \mu_{NZ,9}) \otimes (s_10 = \mu_{Z,10})$$
$$\otimes(s_11 = \mu_{Z,11}) \otimes (s_12 = \mu_{NZ,12})\ then\ \mathbf{f}_7$$

Rule-9:

$$If\ (s_1 = \mu_{Z,1}) \otimes (s_{=}\mu_{Z,2}) \otimes (s_3 = \mu_{NZ,3}) \otimes (s_4 = \mu_{Z,4}) \otimes (s_5 = \mu_{Z,5})$$
$$\otimes(s_6 = \mu_{Z,6}) \otimes (s_7 = \mu_{Z,7}) \otimes (s_8 = \mu_{NZ,8}) \otimes (s_9 = \mu_{Z,9}) \otimes (s_10 = \mu_{NZ,10})$$
$$\otimes(s_11 = \mu_{Z,11}) \otimes (s_12 = \mu_{NZ,12})\ then\ \mathbf{f}_8$$

Rule-10:

$$If\ (s_1 = \mu_{Z,1}) \otimes (s_{=}\mu_{Z,2}) \otimes (s_3 = \mu_{Z,3}) \otimes (s_4 = \mu_{Z,4}) \otimes (s_5 = \mu_{Z,5})$$
$$\otimes(s_6 = \mu_{Z,6}) \otimes (s_7 = \mu_{Z,7}) \otimes (s_8 = \mu_{NZ,8}) \otimes (s_9 = \mu_{NZ,9}) \otimes (s_10 = \mu_{Z,10})$$
$$\otimes(s_11 = \mu_{Z,11}) \otimes (s_12 = \mu_{NZ,12})\ then\ \mathbf{f}_9$$

Rule-11:

$$If\ (s_1 = \mu_{Z,1}) \otimes (s_{=}\mu_{Z,2}) \otimes (s_3 = \mu_{Z,3}) \otimes (s_4 = \mu_{Z,4}) \otimes (s_5 = \mu_{Z,5})$$
$$\otimes(s_6 = \mu_{Z,6}) \otimes (s_7 = \mu_{NZ,7}) \otimes (s_8 = \mu_{NZ,8}) \otimes (s_9 = \mu_{Z,9}) \otimes (s_10 = \mu_{NZ,10})$$
$$\otimes(s_11 = \mu_{Z,11}) \otimes (s_12 = \mu_{Z,12})\ then\ \mathbf{f}_10$$

Rule-12:

$$If\ (s_1 = \mu_{Z,1}) \otimes (s_{=}\mu_{Z,2}) \otimes (s_3 = \mu_{Z,3}) \otimes (s_4 = \mu_{Z,4}) \otimes (s_5 = \mu_{NZ,5})$$
$$\otimes(s_6 = \mu_{Z,6}) \otimes (s_7 = \mu_{Z,7}) \otimes (s_8 = \mu_{Z,8}) \otimes (s_9 = \mu_{Z,9}) \otimes (s_10 = \mu_{Z,10})$$
$$\otimes(s_11 = \mu_{NZ,11}) \otimes (s_12 = \mu_{Z,12})\ then\ \mathbf{f}_11$$

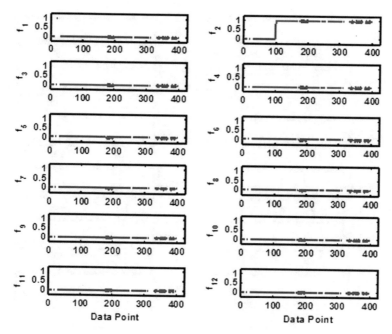

Fig. 7.21 Fault activation level plots of T_{g2} biased with 10% error

Rule-13:

$$If\ (s_1 = \mu_{Z,1}) \otimes (s_{=}\mu_{Z,2}) \otimes (s_3 = \mu_{Z,3}) \otimes (s_4 = \mu_{Z,4}) \otimes (s_5 = \mu_{Z,5})$$
$$\otimes (s_6 = \mu_{Z,6}) \otimes (s_7 = \mu_{NZ,7}) \otimes (s_8 = \mu_{Z,8}) \otimes (s_9 = \mu_{Z,9}) \otimes (s_10 = \mu_{Z,10})$$
$$\otimes (s_11 = \mu_{Z,11}) \otimes (s_12 = \mu_{NZ,12})\ then\ \mathbf{f}_12$$

Now, we first consider the diagnosis of a fault in T_{g1} sensor. Figure 7.21 shows the result of fault diagnosis using the Fuzzy Activation Level (FAL) plots of the test data. It can be observed that the activation level for T_{g2} is much higher than that of other sensors. As such, the diagnosis result points to T_{g2} sensor as the sensor highly likely to carry the bias. Since, we have implanted the fault ourselves, the fault is indeed in T_{g2}. Hence, it can be concluded that the biased sensor, T_{g2}, is diagnosed correctly.

To further explore validity of the method to other sensors too, 10% measurement bias is added to T_{fd2}, T_{fd3} and T_{g4} sensors. Note that the tests are performed under single fault assumption. Plots of FAL for the three test cases are outlined in Figs. 7.22, 7.23 and 7.24, respectively. According to the rules elaborated hereinabove, the plots in the Figures perfectly match the faults. Hence, the results further confirm the high performance of the proposed FDD system.

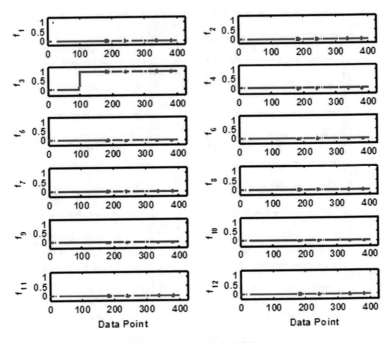

Fig. 7.22 Fault activation level plots of T_{fd2} biased with 10% error

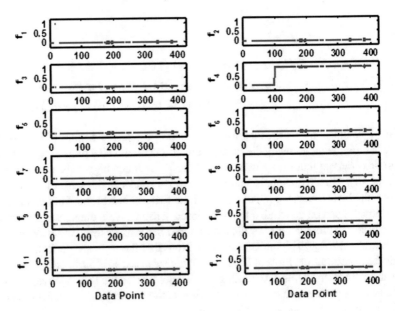

Fig. 7.23 Fault activation level plots of T_{fd3} biased with 10% error

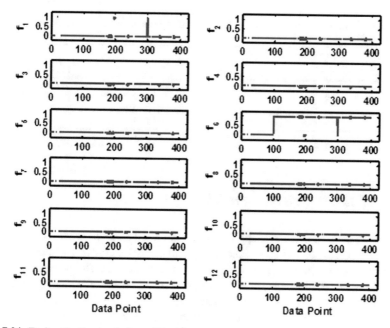

Fig. 7.24 Fault activation level plots of T_{g4} biased with 10% error

7.4 LiBr-H2O Steam Absorption Chiller

Like the GTG and HRSG systems, the faults involved in the SAC are sensor faults, actuator faults and process faults. The SAC works under vacuum pressure. Air leaks to the system may create crystallization and affect performance of the system. The other process related problem is fouling in the cooling water system. This section provides validation test of the FDD scheme using test data for SAC sensor faults. The validation data is generated from normal operation data by adding controlled bias to each sensor measurement.

The models needed for fault detection are the models developed and validated in Sect. 6.3.2. For generating the test data, sensors outside the control loop are considered. These include: chilled water inlet temperature (T_{ch1}), cooling water inlet temperature (T_{co1}), cooling water outlet temperature (T_{co2}), and steam pressure (P_s). A bias error of up to 10% of the design value is added to each measurement. A total of 370 data points is considered for each test case with the addition of the bias being done to data points from 100 to 370. Because each sensor has a bias error, there are a total of four sets of test data. For each test data, residuals are calculated according to the relations listed in Appendix H and Table H.8. In Sect. 7.4.1 is discussed the result from fault detection test while the fault diagnosis is covered in Sect. 7.4.2.

Table 7.12 Minimum detectable abrupt faults, percentage detection and detection delay for SAC sensors: NF based approach

Sensors	Parameter			
	τ_{md}	τ_{tdfp}	τ_{tifp}	τ_{dd} (min)
T_{ch1}	4.9	100	100	84
T_{co1}	5.8	97.78	97.78	184
T_{co2}	4.5	100	100	34
P_s	19.5	94.81(for20%)	94.81	126

7.4.1 Fault Detection in the SAC

The proposed FDD scheme is applied to the generated data and produced test results for the four cases accordingly. Table 7.12 gives the test outcomes. It indicates that all the test cases are detected faulty except the sensor fault related to the steam pressure sensor (P_s). For P_s sensor, the minimum bias that can be detected with true percentage detection reaching 50% is when the added bias is about 19.5% of the design point value (see Tables F.1 and F.2), which is pretty high. For an assumed 10% bias with respect to the design point values, Table 7.12 demonstrates that the added errors are detected with true detection ratio higher than 95%. The corresponding true diagnosis ratio is higher than 95%, which is attractive enough for practical use. An experiment on the minimum bias error that can be detected and diagnosed shows that a bias error of 4.9% in T_{ch1} can be successfully detected by the proposed FDD scheme with a corresponding true detection ratio of 52.96%. The results for the other sensors are also given. The same table contains detection delays if incipient fault at a given rate and maximum value of 10% is assumed. It can be observed that the detection delay is related to the minimum that can be detected. As such, for lower detection percentage it is often the case to witness reduced detection delay. Plots of the residuals for fault free and faulty cases are shown in Figs. 7.25, 7.26, 7.27, 7.28 and 7.29. Figure 7.25 is the residual plot for normal operating condition. As shown by the Figures, the FDD system has discerned the test cases correctly. In the nest section, the diagnosis results are elaborated.

7.4.2 Fault Diagnosis in the SAC

The steps in the diagnosis of faults related to the SAC are no different as compared to the steps followed in the GTG and HRSG. As previously discussed, it pre supposes the calculation of residuals for all measurable parameters (see Appendix F and Table F.9) and the design of a suitable diagnostic matrix. Once that part is complete, the diagnosis follows the signatures in the BDM and the fuzzy rules specifically designed to deal with the sensor faults. The BDM designed for the SAC is diagrammed in

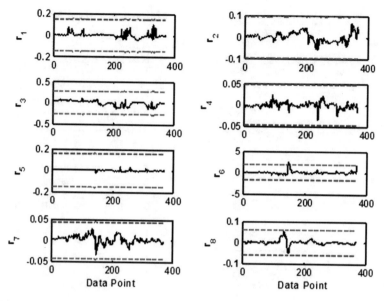

Fig. 7.25   Residuals for normal operating condition no fault case

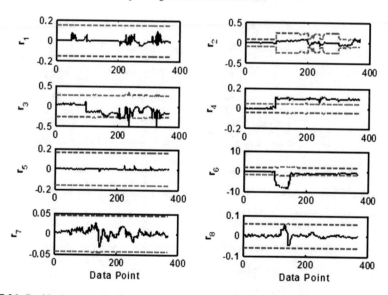

Fig. 7.26   Residuals corresponding to bias error in the T_{ch1} sensor 10% abrupt fault

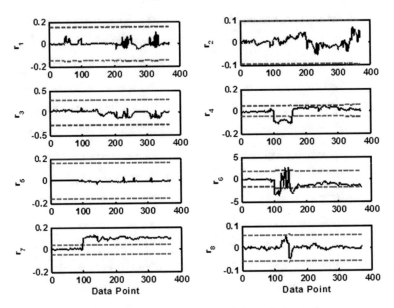

Fig. 7.27 Residuals for bias error in the T_{co1} sensor 10% abrupt fault

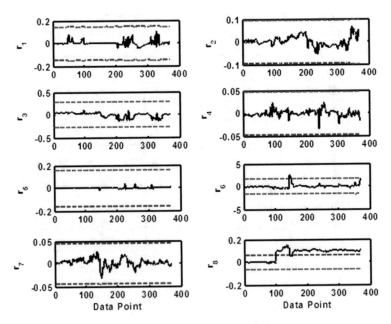

Fig. 7.28 Residuals bias error in the T_{co2} sensor 10% abrupt fault

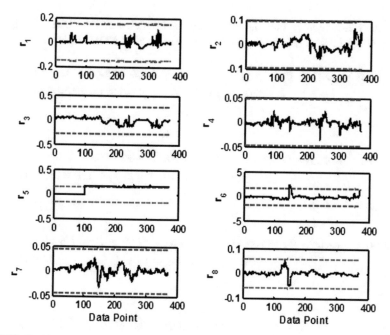

Fig. 7.29 Residuals corresponding to bias error in the P_s sensor 20% abrupt fault

Fig. 7.30 Binary
diagnostics matrix for SAC

R	f_0	f_1	f_2	f_3	f_4	f_5	f_6	f_7	f_8
S_1	1								
S_2	1	1							
S_3		1	1						
S_4				1					
S_5			1		1	1			
S_6						1			
S_7							1		
S_8								1	

Fig. 7.30. A column in the BDM relates to a specific fault in the system. While the first column stands for fault free state, the rest refer to sensor faults. Details about the relationships between fault names and what they refer to can be obtained from Table H.7 in Appendix H.

In the following, we state the fuzzy rules that match with the FDM (see Fig. H.3 in Appendix H) designed on the bases of relations in the BDM. Rule-1:

$$If \ (s_1 = \mu_{Z,1}) \otimes \ (s_2 = \mu_{Z,2}) \otimes (s_3 = \mu_{Z,3}) \otimes (s_4 = \mu_{Z,4}) \otimes (s_5 = \mu_{Z,5})$$
$$\otimes (s_6 = \mu_{Z,6}) \otimes (s_7 = \mu_{Z,7}) \otimes (s_8 = \mu_{Z,8}) \ then \ \mathbf{f}_0$$

Rule-2:

$$If \ (s_1 = \mu_{NZ,1}) \otimes \ (s_2 = \mu_{NZ,2}) \otimes (s_3 = \mu_{Z,3}) \otimes (s_4 = \mu_{Z,4}) \otimes (s_5 = \mu_{Z,5})$$
$$\otimes (s_6 = \mu_{Z,6}) \otimes (s_7 = \mu_{Z,7}) \otimes (s_8 = 0) \ then \ \mathbf{f}_1$$

Rule-3:

$$If \ (s_1 = \mu_{NZ,1}) \otimes \ (s_2 = \mu_{NZ,2}) \otimes (s_3 = \mu_{NZ,3}) \otimes (s_4 = \mu_{NZ,4}) \otimes (s_5 = \mu_{Z,5})$$
$$\otimes (s_6 = \mu_{Z,6}) \otimes (s_7 = \mu_{Z,7}) \otimes (s_8 = \mu_{Z,8}) \ then \ \mathbf{f}_2$$

Rule-4:

$$If \ (s_1 = \mu_{Z,1}) \ \otimes (s_2 = \mu_{Z,2}) \otimes (s_3 = \mu_{NZ,3}) \otimes (s_4 = \mu_{NZ,4}) \otimes (s_5 = \mu_{NZ,5})$$
$$\otimes (s_6 = \mu_{NZ,6}) \otimes (s_7 = 1) \otimes (s_8 = \mu_{Z,8}) \ then \ \mathbf{f}_3$$

Rule-5:

$$If \ (s_1 = \mu_{Z,1}) \otimes \ (s_2 = \mu_{NZ,2}) \otimes (s_3 = \mu_{NZ,3}) \otimes (s_4 = \mu_{NZ,4}) \otimes (s_5 = \mu_{NZ,5})$$
$$\otimes (s_6 = \mu_{NZ,6}) \otimes (s_7 = \mu_{Z,7}) \otimes (s_8 = \mu_{Z,8}) \ then \ \mathbf{f}_4$$

Rule-6:

$$If \ (s_1 = \mu_{Z,1}) \otimes \ (s_2 = \mu_{Z,2}) \otimes (s_3 = \mu_{Z,3}) \otimes (s_4 = \mu_{Z,4}) \otimes (s_5 = \mu_{NZ,5})$$
$$\otimes (s_6 = \mu_{Z,6}) \otimes (s_7 = 0) \otimes (s_8 = \mu_{Z,8}) \ then \ \mathbf{f}_5$$

Rule-7:

$$If \ (s_1 = \mu_{Z,1}) \otimes \ (s_2 = \mu_{Z,2}) \otimes (s_3 = \mu_{Z,3}) \otimes (s_4 = \mu_{NZ,4}) \otimes (s_5 = \mu_{NZ,5})$$
$$\otimes (s_6 = \mu_{NZ,6}) \otimes (s_7 = \mu_{NZ,7}) \otimes (s_8 = \mu_{NZ,8}) \ then \ \mathbf{f}_6$$

Rule-8:

$$If \ (s_1 = \mu_{Z,1}) \otimes \ (s_2 = \mu_{Z,2}) \otimes (s_3 = \mu_{Z,3}) \otimes (s_4 = \mu_{NZ,4}) \otimes (s_5 = \mu_{NZ,5})$$
$$\otimes (s_6 = \mu_{NZ,6}) \otimes (s_7 = \mu_{NZ,7}) \otimes (s_8 = \mu_{Z,8}) \ then \ \mathbf{f}_7$$

Rule-9:

$$\textit{If } (s_1 = \mu_{Z,1})\otimes \ (s_2 = \mu_{Z,2}) \otimes (s_3 = \mu_{Z,3}) \otimes (s_4 = \mu_{Z,4}) \otimes (s_5 = \mu_{Z,5})$$
$$\otimes(s_6 = \mu_{Z,6}) \otimes (s_7 = \mu_{NZ,7}) \otimes (s_8 = \mu_{NZ,8}) \textit{ then } \mathbf{f_8}$$

Now, we apply the BDM and the fuzzy rules to diagnose faults in the three sensors whose detection has been addressed in Sect. 7.4.1. The first sensor that is considered is the sensor for chilled water inlet temperature T_{ch1}. The binary diagnostic signals for the test related to this sensor are shown in Fig. 7.31. According to the BDM designed for the SAC, the fault is in T_{ch1} sensor if the signature is $V_4 = [0\ 0\ 0\ 1\ 0\ 0\ 0\ 0]^T$. As can be seen from Fig. 7.31, this is what exactly happened for data samples including 150 and beyond.

For cooling water temperature sensors T_{co1} and T_{co2}, the binary fault signatures are $V_4 = [0\ 0\ 0\ 0\ 0\ 0\ 1\ 0]^T$ and $V_4 = [0\ 0\ 0\ 0\ 0\ 0\ 0\ 1]^T$, respectively. The use of the signatures in Figs. 7.32 and 7.33 tells that the first graph indicates V_7 while the second features V_4. Since we created the faulty data by implanting percentage bias, the source of change is known from the start. As the final results show, the diagnosis result well agree with the solution. This is a confirmation that the proposed FDD system is indeed capable of successfully detecting and diagnosing sensor faults in the SAC. Similar studies have been conducted in the remaining sensors as well. The conclusion remained similar.

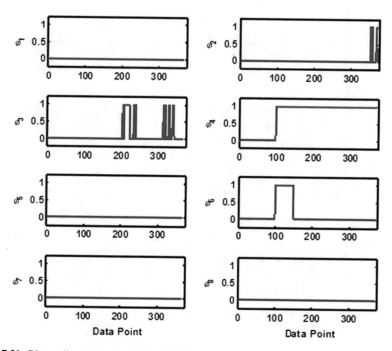

Fig. 7.31 Binary diagnostic signals for a 10% bias in T_{ch1}

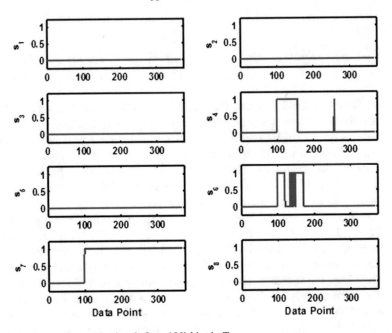

Fig. 7.32 Binary diagnostic signals for a 10% bias in T_{co1}

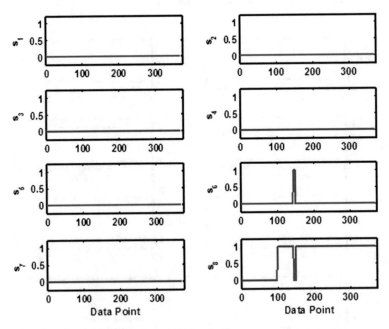

Fig. 7.33 Binary diagnostic signals for a 10% bias in T_{co2}

7.5 Summary

In this chapter, fault detection and diagnosis in the subsystems of the CCP was considered. The FDD system detailed in Chap. 5 was applied with BDM and FDM designed for each subsystem. The capacity of the proposed FDD system was evaluated in terms of minimum sensor bias that can be detected with at least 50% true detection percentage. On the higher side, a 10% bias was assumed and the corresponding true detection ratio and true diagnosis ratio were analysed. To consolidate the result, comparisons were also made between the proposed approach and PCA and AANN methods. The contributions are summarized as follows:

- In Sect. 7.2, the proposed FDD system was applied to detect and diagnose abrupt and incipient sensor faults in the GTG. It covered sensors in the gas path, lube system and generator winding coils. For a 10% sensor bias, the proposed FDD system was found capable of performing with higher than 98% true detection percentage and 100% diagnosis percentage. With respect to PCA and AANN methods, the capacity of NF approach was effective in accurately diagnosing the faults. For sensor fault lower than 10%, the proposed method still capable but the true detection percentage and true diagnosis percentage tends to decrease with the magnitude of fault.
- In Sect. 7.3, the application of the proposed FDD system is demonstrated by considering the HRSG. After designing the BDM and FDM based on several test runs, the method was applied to detect and diagnose sensor faults outside the feedback control loop. Just like the previous test case, the proposed method detected and isolated the assumed faults with higher true detection percentage and true diagnosis percentage. For an assumed average sensor bias of 3%, the method was found capable of detecting the fault with higher than 50% true detection percentage.
- In Sect. 7.4 was explained the test result related to SAC sensor fault detection and diagnosis. Here also, sensors outside the feedback control loop were considered for test. It was found that, for temperature sensors an average sensor bias of 5% can be detected with true detection percentage higher than 50% while a 10% bias can be detected with higher than 97% true detection percentage and 98% true diagnosis percentage.

Finally, it can be concluded that applying PCA, a sensor bias as small as 4.1% in P_2 can be detected with detection percentage more than 50%. The same fault can also be diagnosed with diagnostics percentage higher than 85%.

Chapter 8
Conclusion and Recommendation

8.1 Introduction

This book commenced by presenting the importance of a Fault Detection and Diagnosis (FDD) system in a Cogeneration and Cooling Plant (CCP), and the challenges in designing a suitable FDD system particularly the challenges in the identification of high fidelity models, estimation of model uncertainty and design of diagnostic reasoning algorithms. Survey of literature relevant to the subject revealed that such factors as difficulty in developing a first principle model and the need to pragmatically apply them have led to the use of ubiquitous models linear in the neighbourhood of the design point. It was noted that in addition to the nonlinear characteristics, the most likely scenario that abnormal conditions may occur during machine start-up or in one of the multiple operating regions has not been fully addressed in a way suitable for the end user. Besides, the interconnection between subsystems has been overlooked attributed to most of the research work focusing on testing new algorithms using laboratory scale simulation models rather than investigating the practicality of the method in a wider perspective.

In light of the above mentioned points, the main contribution of this book is in the design of an intelligent fault detection and diagnosis system that takes into account nonlinearities in the system, measured signals from all sources, and interconnection between subsystems. It is also the object of this work to develop a diagnostics framework that allows consideration of uncertainty in the diagnostic result. The task has been accomplished following five major stages (i) study the signals available for measurement and possible fault types; (ii) model identification; (iii) algorithms synthesis for model uncertainty estimation; (iv) design of improved FDD structure; (v) proposing a method to deal with multiple operating regions. In Sect. 8.2 of the chapter, a summary of the conclusion for each stage is documented. In Sect. 8.3, topics that may be considered for further research are outlined.

© Springer International Publishing AG 2018
T.A. Lemma, *A Hybrid Approach for Power Plant Fault Diagnostics*,
Studies in Computational Intelligence 743,
https://doi.org/10.1007/978-3-319-71871-2_8

8.2   Conclusions

Aim of the research has been to find solutions for some of the challenges in the design of fault detection and diagnosis system for a cogeneration and district cooling plant. One of the problems was to devise a method that may take into account nonlinear operating characteristics of the system. To this end, the research fulfilled the requirement by suggesting and extensively validating a modelling technique that is based on NF approach. In the dynamic case, the research also proposed the structuring of NF models in the framework of OBFs. By and large, the testing and validation of the method demonstrated a successful result. The following section summarizes the findings in a lucid manner.

- In addition to the design of an improved FDD system, the first goal was to find a solution to deal with nonlinear characteristics of the system. To this end, the research investigated the use of NF approach trained by four different algorithms. In doing so, while the newly proposed algorithm is fast in training, it failed to retain the transparent structure in the NF model. However, for neural network model specifically for FFM-ANN the method works pretty well for the change in the black-box structure doesn't matter as long as the result is acceptable. Based on multiple test runs, the research focused on using LOLIMOT algorithm for detail design of the proposed FDD system.
- Besides the steady state model, nonlinear dynamic models were constructed in the framework of OBFs. Assuming Laguerre function, a test was also conducted to investigate the effect of pole size and number of Laguerre filters. It was observed that, for the Laguerre based model, the pole size and model order are not critical if LOLIMOT algorithm is used for model training. On the bases of the conclusion, the rest of the analysis concentrated on using NF model under steady state assumption.
- Before designing the FDD system, derivation of uncertainty equations under iid and bounded error assumption, respectively, were also performed. The use of adaptive uncertainty calculation in a LOLIMOT procedure is one contribution from the research. The method was tested prior to applying it in the FDD evaluation.
- After designing the FDD system, the test cases discussed in Chaps. 6 and 7, validated applicability of the method. For most of the faulty cases, the proposed method demonstrated higher true detection percentage and true diagnosis percentage, which makes the method attractive enough for practical use.

Finally, it can be concluded that the proposed ideas and overall FDD design well addressed the main objective and the goals set at the start of the research. However, we still believe that it would be even complete if improvements are made in the areas outlined in the following section.

8.3 Recommendation for Further Research

- Adding an FDD module to deal with the vibration signals.
- Extending the cost estimation modules to include maintenance optimization.
- One major bottleneck in fault diagnosis is the limited number of gas path parameters available for measurement. To accurately tell the status of compressor stages and turbine stage, further studies need to be conducted to identify the optimum number of additional sensors for improved FDD. This idea is also important in terms of the possibility to conduct better exergo-economic estimations and also detail plant reliability analysis.
- Study the possibility of integrating FDD with multi-state reliability predictions and devise sets of strategies on the practical application of the methods in maintenance planning.

Appendix A
Fuzzy Operators

While a fuzzy set A is featured by a membership function $\mu_A : X \to [0, 1]$, there are corresponding operators critical to the application of the concept to system modelling. In the following, some of the fuzzy operators relevant to the thesis are outlined.

A.1 Equality

For all $x \in X$, equality of fuzzy sets dictates that two sets A and B are equal if and only if their membership functions are equal, $\mu_A(x) = \mu_B(x)$.

A.2 Containment

A is said to be contained in B if and only if $\mu_A(x) \leq \mu_B(x)$ for all $x \in X$.

A.3 Complement

For fuzzy sets A and B, B is complement of A if an only if $\mu_B(x) = 1 - \mu_A(x)$ for all $x \in X$. B is commonly denoted as $\bar{A}$ or $\neg A$ or NOT A.

A.4 Intersection

For fuzzy sets A, B and C, C is intersection of A and B if an only if for all $x \in X$ membership degree of C is related to A and B by

© Springer International Publishing AG 2018
T.A. Lemma, *A Hybrid Approach for Power Plant Fault Diagnostics*,
Studies in Computational Intelligence 743,
https://doi.org/10.1007/978-3-319-71871-2

$$\begin{cases} \mu_C(x) = \min(\mu_A(x), \mu_B(x)) \\ \mu_C(x) = \mu_A(x) \wedge \mu_B(x) \\ \mu_C(x) = \mu_A(x) * \mu_B(x) \end{cases}$$

A.5   Union

For fuzzy sets A, B and C, C is the union of A and B if an only if for all $x \in X$ membership degree of C is related to A and B by

$$\begin{cases} \mu_C(x) = \max(\mu_A(x), \mu_B(x)) \\ \mu_C(x) = \mu_A(x) \vee \mu_B(x) \\ \mu_C(x) = \mu_A(x) + \mu_B(x) - \mu_A(x) * \mu_B(x) \end{cases}$$

A.6   Support

The support of fuzzy set A is an ordinary subset of the universe of discourse X that has nonzero membership in A. That is,

$$Support(A) = \{x \in X | \mu_A(x) > 0\}$$

The aforementioned definitions of intersection and union are not the only way to define consistent operations on fuzzy sets. In the following, other definitions of intersection and union will be presented.

A.7   t-Norm or Triangular Norm

It is a binary operation on [0,1] defined by $t : [0, 1] \times [0, 1] \rightarrow [0, 1]$ that satisfies such axioms as

1. Boundary condition: $t(\mu_A(x), 1) = \mu_A(x)$
2. Monotonicity: $t(\mu_A(x), \mu_B(x)) \leq t(\mu_C(x), \mu_D(x))$, if $\mu_A(x) \leq \mu_C(x)$ and $\mu_B(x) \leq \mu_D(x)$
3. Commutativity: $t(\mu_A(x), \mu_B(x)) = t(\mu_B(x), \mu_A(x))$
4. Associativity: $t(\mu_A(x), t(\mu_B(x), \mu_C(x))) = t(t(\mu_A(x), \mu_B(x)), \mu_C(x))$

The four most frequently used t-norm operators are

1. Minimum: $t(\mu_A(x), \mu_B(x)) = \min(\mu_A(x), \mu_B(x)) = \mu_A(x) \wedge \mu_B(x)$
2. Algebraic Product: $t(\mu_A(x), \mu_B(x)) = \mu_A(x).\mu_B(x)$

3. Bounded Difference: $t(\mu_A(x), \mu_B(x)) = \max(0, \mu_A(x) + \mu_B(x) - 1)$
4. Drastic Intersection: $t(\mu_A(x), \mu_B(x)) = \min\{\mu_A(x), \mu_B(x)\}$ if $\mu_A(x) = 1$ or $\mu_B(x) = 1$.

A.8   t-Conorm (s-Norm)

It is also a binary operation of the type $s : [0, 1] \times [0, 1] \rightarrow [0, 1]$ with axioms

1. Boundary condition: $s(\mu_A(x), 0) = \mu_A(x)$
2. Monotonicity: $s(\mu_A(x), \mu_B(x)) \leq s(\mu_C(x), \mu_D(x))$, if $\mu_A(x) \leq \mu_C(x)$ and $\mu_B(x) \leq \mu_D(x)$
3. Commutativity: $s(\mu_A(x), \mu_B(x)) = s(\mu_B(x), \mu_A(x))$
4. Associativity: $s(\mu_A(x), s(\mu_B(x), \mu_C(x))) = s(s(\mu_A(x), \mu_B(x)), \mu_C(x))$

Alternative forms of s-norm operators are

1. Maximum: $s(\mu_A(x), \mu_B(x)) = \max(\mu_A(x), \mu_B(x))$
2. Algebraic Sum: $s(\mu_A(x), \mu_B(x)) = \mu_A(x) + \mu_B(x) - \mu_A(x).\mu_B(x)$
3. Bounded Sum: $s(\mu_A(x), \mu_B(x)) = \min\{1, \mu_A(x) + \mu_B(x)\}$
4. Drastic Sum: $s(\mu_A(x), \mu_B(x)) = \max\{\mu_A(x), \mu_B(x)\}$ if $\mu_A(x) = 0$ or $\mu_B(x) = 0$.

A.9   Cartesian Product

Given fuzzy sets A and B in X and Y, respectively, the Cartesian product R of A and B, is a fuzzy set in the product space $X \times Y$ with the membership function expressed as

$$\mu_{A \times B}(x, y) = \min\{\mu_A(x).\mu_B(x)\}.$$

A.10   Cartesian Co-product

For fuzzy sets A and B, their Cartesian product designated as $A + B$ is a fuzzy set in the product space $X \times Y$ with the membership function given by

$$\mu_{A \times B}(x, y) = \max\{\mu_A(x).\mu_B(x)\}.$$

Appendix B
First Order Derivatives of Neural Network Based Output Functions

B.1 Feed Forward Multilayer Artificial Neural Network

The output from a FFM-ANN is described by (3.3). The first derivative of y^p with respect to the weights in the second layer $\tilde{\boldsymbol{\theta}} = [\theta_0\ \theta_1\ \dots\ \theta_{nk}]^T$ is given by:

$$\frac{\partial y^p}{\partial \boldsymbol{\theta}} = \begin{bmatrix} 1 & \cdots & f_k^p & \cdots & f_{nk}^p \end{bmatrix}^T = \tilde{\mathbf{f}}^T \tag{B.1}$$

And, assuming hyperbolic tangent function as an activation function, the derivative of y^p with respect to the weights in first layer are:

$$\frac{\partial y^p}{\partial \mathbf{W}^{<1>}} = diag([1\ \mathbf{u}^p])\mathbf{\Gamma}_{(nu+1,nk)}diag(1 - (f_k^p)^2) * diag(\boldsymbol{\theta}) \tag{B.2}$$

where, $\boldsymbol{\theta} = [\theta_1\ \theta_2\ \dots\ \theta_{nk}]^T$.

Or, for any form of activation function,

$$\frac{\partial y^p}{\partial \mathbf{W}^{<1>}} = diag([1\ \mathbf{u}^p])\mathbf{\Gamma}_{(nu+1,nk)}diag((f_k^p)') * diag(\boldsymbol{\theta}) \tag{B.3}$$

In the stated equations, *diag*: stands for diagonal matrix form of a vector, and $(f_k^p)'$ is first derivative of the f_k^p with respect to the input to the same function. And,

$$\mathbf{\Gamma}_{(nu+1,nk)} = \begin{bmatrix} 1 & 1 & \cdots & 1 \\ 1 & 1 & \cdots & 1 \\ \vdots & \vdots & \ddots & \vdots \\ 1 & 1 & \cdots & 1 \end{bmatrix}_{(nu+1,nk)}$$

© Springer International Publishing AG 2018
T.A. Lemma, *A Hybrid Approach for Power Plant Fault Diagnostics*,
Studies in Computational Intelligence 743,
https://doi.org/10.1007/978-3-319-71871-2

In estimating confidence interval of the feed forward multilayer neural network, (B.1) and (B.2) are combined to form the Jacobian matrix (Refer to Chap. 4 and Sect. 4.2.2).

B.2 Radial Basis Function Artificial Neural Network

Structure of the derivative of the output from RBF-ANN with respect to the second layer weights is similar to the one derived for FFM- ANN (B.1). For the weights in the first layer, it is common practice to cluster the date prior to estimating the weights in the second layer. In the following, however, it is assumed that the model uncertainty is affected by all model parameters.

$$\frac{\partial y^p}{\partial \tilde{\theta}} = \begin{bmatrix} 1 & \cdots & f_k^p & \cdots & f_{nk}^p \end{bmatrix}^T = \tilde{f}^T \tag{B.4}$$

The assumption of single output still holds. Derivative of y^p with respect to 2nd layer parameters takes the same for as the one derived for MLP-NN model. For the derivative of y^p with respect to 1st layer σ_k's,

$$\frac{\partial f^p}{\partial \Sigma} = \begin{bmatrix} \frac{f_1^p}{\sigma_1^3} \left(\frac{\partial y^p}{\partial f_1^p} \right) \|u^p - c_1\|^2 & \frac{f_k^p}{\sigma_k^3} \left(\frac{\partial y^p}{\partial f_k^p} \right) \|u^p - c_k\|^2 & \frac{f_{nk}^p}{\sigma_{nk}^3} \left(\frac{\partial y^p}{\partial f_{nk}^p} \right) \|u^p - c_{nk}\|^2 \end{bmatrix}$$

Which can be written in short form as,

$$\frac{\partial y^p}{\partial \Sigma} = diag \left(\frac{\partial y^p}{\partial f^p} \right) * diag \left(f^p \right) * S_{ICU} * D_{SQ} \tag{B.5}$$

where,

$$\frac{\partial y^p}{\partial f^p} = \begin{bmatrix} \theta_1 & \theta_k & \theta_{nk} \end{bmatrix},$$

$$S_{ICU} = diag \left[\left(\sigma_1^{-3} \ \sigma_k^{-3} \ \sigma_{nk}^{-3} \right) \right], \text{ and}$$

$$D_{SQ} = diag \left(\begin{bmatrix} \|u^p - c_1\|^2 & \|u^p - c_k\|^2 & \|u^p - c_{nk}\|^2 \end{bmatrix} \right)$$

Derivatives of the network output y^p with respect to $c_{j,k}$'s:

$$\frac{\partial y^p}{\partial C} = \begin{bmatrix} \frac{f_1^p}{\sigma_1^2} \left(\frac{\partial y^p}{\partial f_1} \right) \begin{bmatrix} u_1^p - c_{1,1} \\ u_j^p - c_{j,1} \\ u_{nu}^p - c_{nu,1} \end{bmatrix} & \frac{f_k^p}{\sigma_k^2} \left(\frac{\partial y^p}{\partial f_k} \right) \begin{bmatrix} u_1^p - c_{1,k} \\ u_j^p - c_{j,k} \\ u_{nu}^p - c_{nu,k} \end{bmatrix} & \frac{f_{nk}^p}{\sigma_{nk}^2} \left(\frac{\partial y^p}{\partial f_{nk}} \right) \begin{bmatrix} u_1^p - c_{1,nk} \\ u_j^p - c_{j,nk} \\ u_{nu}^p - c_{nu,nk} \end{bmatrix} \end{bmatrix}$$

Note the tame subtraction operator between a vector and a matrix. Short form of the equation will be

$$\frac{\partial y^P}{\partial \mathbf{C}} = (\mathbf{u}^P . - \mathbf{C}) * diag \left(\frac{\partial y^P}{\partial \mathbf{f}^P} \right) * diag \left(\mathbf{f}^P \right) * \left(\mathbf{S}_{ISQ} \right) \qquad \text{(B.6)}$$

B.3 Normalized Radial Basis Function Artificial Neural Network

In NRBF-ANN, there is offset term in the second layer. Hence, the derivative of the output with respect to the parameters in the output layer is

$$\frac{\partial y^P}{\partial \boldsymbol{\theta}} = \left[\varphi_1^P \cdots \varphi_k^P \cdots \varphi_{nk}^P \right]^T \qquad \text{(B.7)}$$

where,

$$\varphi_k^P = \frac{f_k^P}{\sum_{k=1}^{nk} f_k^P}.$$

Assuming that $f_\Sigma = \sum_{k=1}^{nk} f_k^P$, taking derivative of the network output with respect to f^P yields

$$\frac{\partial y^P}{\partial \mathbf{f}^P} = \frac{1}{f_\Sigma^2} . * \begin{bmatrix} f_\Sigma - f_1^P & f_k^P & f_{nk}^P \\ f_1^P & f_\Sigma - f_k^P & f_{nk}^P \\ f_1^P & f_k^P & f_\Sigma - f_{nk}^P \end{bmatrix} \begin{bmatrix} \theta_1 \\ \theta_k \\ \theta_{nk} \end{bmatrix}$$

The corresponding short form will be

$$\frac{\partial y^P}{\partial \mathbf{f}^P} = \frac{1}{f_\Sigma^2} . * \left(f_\Sigma . * \mathbf{I}_{(nk,nk)} - \Gamma_{(nk,nk)} * diag \left(\mathbf{f}^P \right) \right) * \theta$$

Derivative f^P of with respect to the spread terms:

$$\frac{\partial \mathbf{f}^P}{\partial \Sigma} = \left[\frac{f_1^P}{\sigma_1^3} \| \mathbf{u}^P - c_1 \|^2 \ \frac{f_k^P}{\sigma_k^3} \| \mathbf{u}^P - c_k \|^2 \ \frac{f_{nk}^P}{\sigma_{nk}^3} \| \mathbf{u}^P - c_{nk} \|^2 \right]$$

Based on this result,

$$\frac{\partial y^P}{\partial \Sigma} = diag \left(\frac{\partial y^P}{\partial \mathbf{f}^P} \right) * diag \left(\mathbf{f}^P \right) * \mathbf{S}_{ICU} * \mathbf{D}_{SQ} \qquad \text{(B.8)}$$

where,
$$\mathbf{S}_{ICU} = diag \left[\left(\sigma_1^{-3} \ \sigma_k^{-3} \ \sigma_{nk}^{-3} \right) \right], \text{ and}$$

$$\mathbf{D}_{SQ} = diag\left(\left[\; \|u^p - c_1\|^2 \;\; \|u^p - c_k\|^2 \;\; \|u^p - c_{nk}\|^2 \;\right]\right)$$

The derivative network output with respect to centre matrix in the first layer:

$$\frac{\partial y^p}{\partial \mathbf{C}} = \left[\; \frac{f_1^p}{\sigma_1^2}\left(\frac{\partial y^p}{\partial f_1}\right) \begin{bmatrix} u_1^p - c_{1,1} \\ u_j^p - c_{j,1} \\ u_{nu}^p - c_{nu,1} \end{bmatrix} \; \frac{f_k^p}{\sigma_k^2}\left(\frac{\partial y^p}{\partial f_k}\right) \begin{bmatrix} u_1^p - c_{1,k} \\ u_j^p - c_{j,k} \\ u_{nu}^p - c_{nu,k} \end{bmatrix} \; \frac{f_{nk}^p}{\sigma_{nk}^2}\left(\frac{\partial y^p}{\partial f_{nk}}\right) \begin{bmatrix} u_1^p - c_{1,nk} \\ u_j^p - c_{j,nk} \\ u_{nu}^p - c_{nu,nk} \end{bmatrix} \;\right]$$

Short form,

$$\frac{\partial y^p}{\partial \mathbf{C}} = \left(u^p . - \mathbf{C}\right) * diag\left(\frac{\partial y^p}{\partial f^p}\right) * diag\left(f^p\right) * \left(\mathbf{S}_{ISQ}\right) \tag{B.9}$$

B.4 TSK Based Fuzzy Model

There are three sources of model parameters in the TSK based NF model. For the derivative of y^p with respect to $\theta_{j,k}$'s, which is from 5th layer:

$$\frac{\partial y^p}{\partial \widetilde{\mathbf{W}}^{<5>}} = \left[\; \varphi_1^p \begin{bmatrix} 1 \\ u_1 \\ u_j \\ u_{nu} \end{bmatrix} \; \varphi_k^p \begin{bmatrix} 1 \\ u_1 \\ u_j \\ u_{nu} \end{bmatrix} \; \varphi_{nk}^p \begin{bmatrix} 1 \\ u_1 \\ u_j \\ u_{nu} \end{bmatrix} \;\right]$$

Short form

$$\frac{\partial y^p}{\partial \widetilde{W}^{<5>}} = diag\left(\tilde{u}^p\right) * \Gamma_{(nu+1,nk)} * diag\left(\varphi_1^p\right) \tag{B.10}$$

Calculation of the derivative of y^p with respect to $\sigma_k's$ demands the derivative of y^p with respect to α^p. Hence,

$$\frac{\partial y^p}{\partial p} = \frac{1}{\alpha_\Sigma^2} . * \begin{bmatrix} \alpha_\Sigma - \alpha_1^p & \alpha_k^p & \alpha_{nk}^p \\ \alpha_1^p & \alpha_\Sigma - \alpha_k^p & \alpha_{nk}^p \\ \alpha_1^p & \alpha_k^p & \alpha_\Sigma - \alpha_{nk}^p \end{bmatrix} \begin{bmatrix} z_1^p \\ z_k^p \\ z_{nc}^p \end{bmatrix}$$

where $\alpha_\Sigma = \left(\sum_{k=1}^{nk} \alpha_k^p\right)$
Short form,

$$\frac{\partial y^p}{\partial \alpha^p} = \frac{1}{\alpha_\Sigma^2} . * \left(\Sigma . * \mathbf{I}_{(nk,nk)} - \Gamma_{(nk,nk)} * diag\left(\alpha^p\right)\right) * \left(z^p\right)^T \tag{B.11}$$

To use this formalism,

$$\frac{\partial y^p}{\partial \Sigma} = \begin{bmatrix} \frac{\alpha_1^p \|u^p - c_1\|^2}{\sigma_1^3} \left(\frac{\partial y^p}{\partial \alpha_1} \right) \\ \frac{\alpha_k^p \|u^p - c_k\|^2}{\sigma_k^3} \left(\frac{\partial y^p}{\partial \alpha_k} \right) \\ \frac{\alpha_{nk}^p \|u^p - c_{nk}\|^2}{\sigma_{nk}^3} \left(\frac{\partial y^p}{\partial \alpha_{nk}} \right) \end{bmatrix}$$

In short form,

$$\frac{\partial y^p}{\partial \Sigma} = diag \left(\frac{\partial y^p}{\partial p} \right) * diag \left(^p \right) * S_{ICU} * D_{SQ} \tag{B.12}$$

where,

$$S_{ICU} = diag \left[\sigma_1^{-3} \ \sigma_k^{-3} \ \sigma_{nk}^{-3} \right], \text{ and}$$

$$D_{SQ} = diag \left(\left[\ \|u^p - c_1\|^2 \ \|u^p - c_k\|^2 \ \|u^p - c_{nk}\|^2 \ \right] \right)$$

Note that the result from (B.12) is a diagonal matrix. Hence, to put it in column vector form, it should be written as

$$\frac{\partial y^p}{\partial \Sigma} = diag \left(diag \left(\frac{\partial y^p}{\partial p} \right) * diag \left(^p \right) * S_{ICU} * D_{SQ} \right)$$

As for the derivative of y^p with respect to the membership center matrix, C:

$$\frac{\partial y^p}{\partial C} = \begin{bmatrix} \frac{\alpha_1^p}{\sigma_1^2} \left(\frac{\partial y^p}{\partial \alpha_1} \right) \begin{bmatrix} u_1^p - c_{1,1} \\ u_j^p - c_{j,1} \\ u_{nu}^p - c_{nu,1} \end{bmatrix} \frac{\alpha_k^p}{\sigma_k^2} \left(\frac{\partial y^p}{\partial \alpha_k} \right) \begin{bmatrix} u_1^p - c_{1,k} \\ u_j^p - c_{j,k} \\ u_{nu}^p - c_{nu,k} \end{bmatrix} \frac{\alpha_{nk}^p}{\sigma_{nk}^2} \left(\frac{\partial y^p}{\partial \alpha_{nk}} \right) \begin{bmatrix} u_1^p - c_{1,nk} \\ u_j^p - c_{j,nk} \\ u_{nu}^p - c_{nu,nk} \end{bmatrix} \end{bmatrix}$$

Writing it in short form,

$$\frac{\partial y^p}{\partial C} = \left(u^p . - C \right) * diag \left(\frac{\partial y^p}{\partial p} \right) * diag \left(^p \right) * \left(S_{ISQ} \right) \tag{B.13}$$

where

$$S_{ISQ} = diag \left[\left(\sigma_1^{-2} \ \sigma_k^{-2} \ \sigma_{nk}^{-2} \right) \right].$$

Not the tame subtraction ".−" operator between a column vector and a matrix.

B.5 Generalized Orthonormal Basis Functions

In this section, the matrices formed for a GOBF based model are given. While there are possibilities to consider a mix of real and complex poles, in the following only the case where all the poles are real and different from one another is considered.

Assuming that $\xi = \begin{bmatrix} \xi_1 & \xi_2 & \cdots & \xi_m \end{bmatrix}$ and $\beta_j = \sqrt{1 - |\xi_j|^2}$, the matrices $\Phi\,(\xi)$ and $\Gamma\,(\xi)$ are given by (B.14) and (B.15), respectively.

$$\Phi\,(\xi) = [\Xi_1\,(\xi)]^{-1}\,\Xi_2\,(\xi) \tag{B.14}$$

$$\Gamma\,(\xi) = [\Xi_1\,(\xi)]^{-1}\begin{bmatrix} 1 & 0 & 0 & \cdots & 0 \end{bmatrix}^T \tag{B.15}$$

where,

$$\Xi_1\,(\xi) = \begin{bmatrix} 1 & 0 & 0 & 0 & 0 \\ \xi_1\left(\frac{\beta_2}{\beta_1}\right) & 1 & 0 & 0 & 0 \\ 0 & \xi_2\left(\frac{\beta_3}{\beta_2}\right) & 1 & \cdots & \vdots \\ \vdots & \vdots & \cdots & \ddots & 0 \\ 0 & 0 & \cdots & \xi_{m-1}\left(\frac{\beta_m}{\beta_{m-1}}\right) & 1 \end{bmatrix},$$

$$\Xi_2\,(\xi) = \begin{bmatrix} \xi_1 & 0 & 0 & 0 & 0 \\ \left(\frac{\beta_2}{\beta_1}\right) & \xi_2 & 0 & 0 & 0 \\ 0 & \left(\frac{\beta_3}{\beta_2}\right) & \xi_3 & \cdots & \vdots \\ \vdots & \vdots & \cdots & \ddots & 0 \\ 0 & 0 & \cdots & \left(\frac{\beta_m}{\beta_{m-1}}\right) & \xi_m \end{bmatrix}$$

In the case where all the poles are identical, the GOBF reduces to Laguerre model.

B.6 Calculation of Time Delay, Model Order and Time Constant

For the calculation of the time delay T_d, model order n, and time constant T, the higher order transfer function as given by (B.16) can be assumed. In the first trial, the time delay is deemed zero, which stands for ideal case.

$$G(s) = \frac{A}{(Ts + 1)^n}e^{-T_d s} \tag{B.16}$$

where, A is the process gain. In the classical method, a graphical approach is used to estimate the time delay based on the data at the inflection point as shown in Fig. B.1 [1, 2]. However, so as to avoid manual calculation, the slop at the inflection point can be estimated by using (B.17) [2, 3].

$$\alpha_i = \frac{(y_{p+1} - y_p) + \Delta y_{max}}{\Delta t_p} \tag{B.17}$$

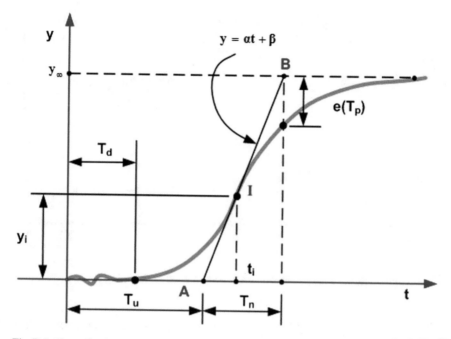

Fig. B.1 Normalized step response of a higher order system

where, $\Delta y_{\max} = \max\{\Delta y_p = y_p - y_{p-1}, p = 1, 2, \ldots, N_d\}$ and $\Delta t_p = t_{p+1} - t_{p-1}$. The line tangent to the step response curve and at the inflection point is defined as

$$y = \alpha t + \beta \tag{B.18}$$

where, α is the slop and β is the y intercept. The line passes through points A, I and B. Using the coordinates at points A and B, the constants in the linear equation are $\alpha = y_\infty/T_n$ and $\beta = -\alpha T_u$. The step response of the transfer function G(s) with $T_d = 0$ is obtained by inverse Laplace transform and is given by

$$y(t) = 1 - e^{-t/T} \sum_{k=0}^{n-1} \frac{1}{k!}\left(\frac{t}{T}\right)^k \tag{B.19}$$

And, the first and second derivatives are:

$$\dot{y}(t) = \frac{1}{T^n(n-1)!}t^{n-1}e^{-t/T}$$

$$\ddot{y}(t) = \frac{1}{T^n}\left[\frac{t^{n-2}}{(n-2)!} - \frac{1}{T}\frac{t^{n-1}}{(n-1)!}\right]e^{-t/T}$$

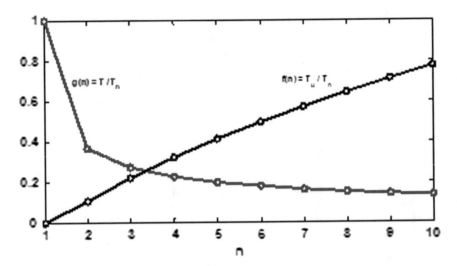

Fig. B.2 Plots relevant to the calculation of apparent time and contribution time

The evaluation of these equations at the inflection point and axis crossing points leads to:

$$t_i = T(n-1)$$

$$\frac{T}{T_n} = \frac{(n-1)}{(n-1)!} e^{-(n-1)} \tag{B.20}$$

$$\frac{T_u}{T_n} = e^{-(n-1)} \left[\frac{(n-1)^n}{(n-1)!} + \sum_{k=0}^{n-1} \frac{1}{k!}(n-1)^k \right] - 1 \tag{B.21}$$

Equations (B.20) and (B.21) are functions of only the model order. Figure B.2 shows plots of the two equations. Making use of the equations or the graphs, model identification can be performed according to the following procedure.

1. Calculate the slop of the tangent line from (B.17) and (B.18)
2. Calculate T_u from (B.18) and data at the inflection point. As such, $T_u = \Delta t_p \left[\frac{y_i}{\Delta y_{p,max}} - \frac{t_i}{\Delta t_p} \right]$. Where, $\Delta t_p = (t_p - t_{p-1})|_{\Delta y_{p,max}}$
3. Calculate T_n form (B.18) and data at point B in Fig. B.1.

$$T_n = \Delta t_p \left[\frac{y_\infty - y_i}{\Delta y_{p,max}} - \frac{t_i}{\Delta t_p} \right] - T_u$$

4. Calculate $f_{real} = T_u / T_n$ and use Fig. B.2 to get approximate value of n with the condition that $f(n) \leq f_{real} \leq f(n+1)$.

5. Calculate the time delay as the difference between the real and theoretical time T_u. That is,

$$T_d = [f_{real} - f(n)].T_n$$

6. Use the estimated value of n to calculate for the time constant.

$$T = T_n.g(n)$$

Signifying the sampling time by Ts, the dominant pole can be approximated by:

$$\xi = \exp(-Ts/T) \tag{B.22}$$

References

1. Seborg, D. E., Edgar, T. F., & Mellichamp, D. A. (2003). *Process dynamics and control*. John Wiley and Sons.
2. Mikles, J. & Fikar, M. (2007). *Process modelling identification and control*. Springer-Verlag.
3. Patwardhan, S. C., & Shah, S. L. (2005). From data to diagnosis and control using generalized orthonormal basis filters. Part I: Development of state observers. *Journal of Process Control, 15*, 819–835.

Appendix C
Data Pre-processing Techniques and Model Performance Measures

C.1 Data Pre-processing

In model identification, data from measurements are used for model training as well as model validation. Common problems in measured data are they may be corrupted by measurement noise or outliers. In some cases, there could also be data figure size differences and long data set. Direct use of a data having the said features spoils accuracy of the model and there may also be convergence problem during model training. Hence appropriate techniques need to be used to circumvent problems. In the subsequent sections, the approaches commonly used by model identification and FDD community will be considered.

C.1.1 Measurement Noise and Outlier Removal

The signals available for measurement are contaminated by noise and outliers. In fault detection and diagnosis, this condition leads to false alarms. Among the method suitable for reducing their effect are: moving average, rank permutation, wavelet analysis [1], etc.

Wavelet analysis is a variable window transformation that uses short and long time intervals to analyze high frequency and low frequency components, respectively, of a signal. As compared to other transformation techniques, wavelet preserves the time and frequency component of a signal. There are two forms of this transformation: Continuous Wavelet Transform (CWT) and Discrete Wavelet Transform (DWT). The continuous wavelet transform of a signal $y(t)$-assumed square integrable - is the convolution of the signal with $\psi_{a,b}(t) \in L^2(R)$ and integrated over the interval $[-\infty, \infty]$, (C.1).

$$W_y(a, b) = \langle y(t), \psi_{a,b}(t) \rangle = \int_{-\infty}^{\infty} y(t)\psi_{a,b}^*(t)dt \qquad (C.1)$$

© Springer International Publishing AG 2018
T.A. Lemma, *A Hybrid Approach for Power Plant Fault Diagnostics*,
Studies in Computational Intelligence 743,
https://doi.org/10.1007/978-3-319-71871-2

The original signal can be constructed by considering:

$$y(t) = \frac{1}{W_\psi} \int_{-\infty}^{\infty} \left\{ \int_{-\infty}^{\infty} \langle W_y(a, b), y(t) \rangle \psi_{a,b}(t) db \right\} \frac{da}{a^2} \qquad (C.2)$$

where, W_ψ is a constant depending up on the basis function $\psi(t)$. The basis functions $\psi_{a,b}(t) \in L^2(R)$ satisfy orthogonality condition and are generated from a mother wavelet function $\psi(t)$, Fig. C.1, by scaling and shifting mechanisms. $\psi *_{a,b}(t)$ is the complex conjugate of $\psi_{a,b}(t)$. That is,

$$\psi_{a,b}(t) = \frac{1}{\sqrt{a}} \psi \left(\frac{t-b}{a} \right), \quad a \in R^+ - \{0\}, \quad b \in R \qquad (C.3)$$

In (C.3), a is the scaling parameter and b is the shifting parameter. These parameters could be considered either as continuous or discrete. But, the common and suitable form for computational efficiency reasons is to assume that both are discretized dyadically by a factor of 2^i for scaling and and $2^i j$ for shifting; where i and j are integers. Under these assumptions, (C.3) changes to

$$W_f(a, b) = 2^{-i/2} \int y(t) \psi_{i,j}^* (2^{-it} - j) dt \qquad (C.4)$$

As compared to CWT, DWT governed by (C.4) is found to yield fast computations. In estimating DWT, the original signal $y(t)$ is filtered by low pass filters L_0 and high pass filters H_0 in a successive way until the required level is reached, Fig. C.2. In the inverse DWT, the original signal is synthesized from the decomposed signals according to the structure indicated in Fig. C.3.

$$y(t) = \sum_{j \in N} \sum_{k \in N} W(j, k) \psi_{j,k}(t) \qquad (C.5)$$

Detail signal at level L is:

$$D(t) = \sum_{k \in N} W(j, k) \psi_{j,k}(t)$$

And the approximation at level L is:

$$A_{J-1} = \sum_{k \in N} D_j$$

In terms of detail and approximate signals, the original signal is calculated as

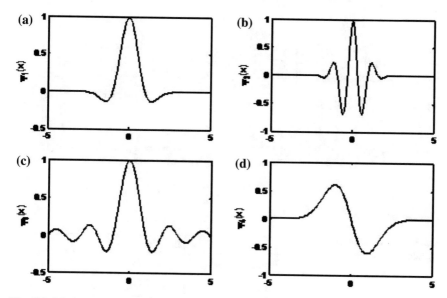

Fig. C.1 Mother wavelets: **a** Mexican Hat; **b** Morlet; **c** Shanon; **d** 1st Derivative of Gaussian

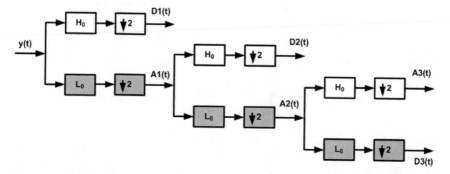

Fig. C.2 Structure of wavelet decomposition

$$y(t) = A_j + \sum_{k \in N} D_j$$

C.1.2 Data Normalization

Model identification using computational intelligence techniques requires that data is normalized before used for training, validation or testing. This is done to avoid one figure dominating the other in the modeling process. The common approach is

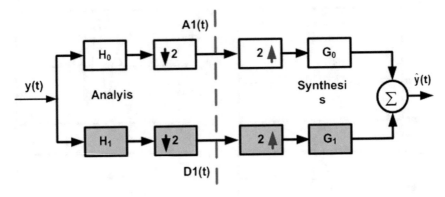

Fig. C.3 Structure of wavelet reconstruction

Table C.1 Normalization equations

Parameter	Equation
Normalized data	$\zeta_{p,j} = \frac{x_{p,j} - \mu_j}{\sigma_j}$
Mean	$\mu_j = \frac{1}{N_d} \sum_{p=1}^{N_d} x_{p,j}$
Standard deviation	$\sigma_j = \left[\frac{1}{N_d} \sum_{p=1}^{N_d} (x_{p,j} - \mu_j)^2 \right]^{\frac{1}{2}}$

to scale the data in a certain range applying (C.6).

$$\zeta_{p,j} = \zeta_{\min} + \left(\frac{\zeta_{\max} - \zeta_{\min}}{x_{j,\max} - x_{j,\min}} \right) (x_{p,j} - x_{j,\min}) \tag{C.6}$$

where, $p = 1, \ldots, N_d$, $j = 1, \ldots, nx$, $x_{\min}$ is the minimum and $x_{\max}$ is the maximum values, respectively, of the data to be normalized. $\zeta_{\min}$ and $\zeta_{\max}$ indicate the minimum and maximum, respectively, of the normalizing data. In a typical case $\zeta_{\min}$ is 0 and $\zeta_{\max}$ is 1.

A data can also be normalized around mean and standard deviation calculated for each variable in the data set. This approach contains the data in a set characterized by zero means and standard deviation of 1, Fig. D.1. The relevant equations are given in Table C.1.

C.2   Data Dimension Reduction

One difficulty in NF based model identification is dimension damnation. As such, increase in number of inputs results in a need for excessive amount of data. Even then, such an approach may end up with difficulties in training the model. There are many dimension reduction techniques. Among them is principal component analysis, nonlinear principal component analysis, and independent component analysis. The following section presents the concepts behind principal component analysis.

C.2.1   Principal Component Analysis

In the PCA method [2], a data matrix $\mathbf{Z} \in R^{Nd \times (n_x + n_y)}$ is projected to a lower dimensional space that describes most of the variance in the data set.

$$\mathbf{Z} = \sum_{i=1}^{k} \mathbf{t}_i \phi_i^T + \mathbf{E} \tag{C.7}$$

where, $\mathbf{E}$ is the residual matrix, ϕ_i is the loading vector – eigenvectors of the covariance matrix $\mathbf{Z}^T \mathbf{Z}/(Nd - 1)$–and $\mathbf{t}_i$ is the projection of $\mathbf{Z}$ in the direction of ϕ_i. The number of loading vectors k considered for the projection is determined either by proportion of trace explained method or SCREE test [2].

The variance explained by a principal component is the ratio of the eigenvalue and the total variance in the original data.

$$\%Var_j = \frac{\lambda_j}{\sum_{j=1}^{l} \lambda_j} \times 100\% \tag{C.8}$$

The cumulative variance explained by the first j principal components is given by

$$\%CumVar_j = \frac{\sum_{k=1}^{j} \lambda_k}{\sum_{k=1}^{l} \lambda_k} \times 100\% \tag{C.9}$$

The fit between a data point and the reference model is calculated as

$$\mathbf{e}_i = \mathbf{Z}_i - \hat{\mathbf{Z}}_i \tag{C.10}$$

where, e_i is the i-th row of Z_i. While Q-statistic $e_i = Z_i$ is used to measure how well a sample fits to the PCA model, Hotelling's T^2 is employed to indicate the distance of an estimated sample from the multivariate mean of the data. The T^2 is given as:

$$T^2 = \sum_{i=1}^{k} (t_i / \lambda_i)^2 \tag{C.11}$$

Calculation of Model Threshold for PCA

For principal component analysis, the threshold limit for Q-statistics [2] is formulated as

$$Q_{th} = \theta_1 \left[1 + \frac{C_\alpha \sqrt{2\theta_2 h_o^2}}{\theta_1} + \frac{\theta_2 h_o (h_o - 1)}{\theta_1^2} \right]^{\frac{1}{2}} \tag{C.12}$$

where, $\theta_i = \sum_{j=k+1}^{m} \lambda_j^i (i = 1, 2, 3)$, C_α is the confidence limit for $(1-\alpha)$ percentage in a normal distribution and $h_o = 1 - 2\theta_1\theta_3/(3\theta_2^2)$. As for T^2, the threshold is given as

$$T_m^2 = \frac{m(N_d - 1)(N_d + 1)}{N_d(N_d - m)} F_\alpha(m, N_d - m) \tag{C.13}$$

where, $F_\alpha(m, N_d - m)$ is the upper $(1 - \alpha)$ percentage point of F distribution with m and $(N_d - m)$ degrees of freedom.

C.2.2 *Auto-associative Neural Network*

As Kramer [3] suggested, nonlinear principal component analysis based on AANN approach could also be employed to extract correlations in a data. We have included AANN with the intent that it will be used for comparison purpose while there is fault detection and diagnosis by other techniques. In the following, the main equations are derived with the assumption that there are $nx \in N$ measurable input variables, $ny \in N$ measurable output variables and $N_d \in N$ as size of reading for each variable. Unless stated, $j \in \{1, \ldots, nx\}$ is the index for input variable x, $q \in \{1, \ldots, ny\}$ is the index for output variable y and $p \in \{1, \ldots, ny\}$ is the index for a single data set. Outputs from the nodes in the layer-1, layer-2 and layer-3 are signified by α_l^p, Z_k^p and ϕ_m^p, respectively. Structure of the network is shown in Fig. C.4.

Each layer output equations are derived based on a similar assumption that number of data is N_d. nx, nl, nc, nm and ny signify number of input variables, nodes in the second layer, nodes in the bottle-neck layer, nodes in the fourth, nodes in the output variables, respectively. Because AANN is a self-mapping network, nx and ny are equal. Formal relations can be written for the outputs from each layer ($\mathbf{A}_{(Nd,nl)}$, $\mathbf{Z}_{(Nd,nc)}$, $\mathbf{\Phi}_{(Nd,nm)}$, and $\mathbf{Y}_{(Nd,ny)}$) after stacked over the whole data set. In Table C.2 are listed the relations at each stage. Linear activation functions are used in all nodes except for nodes in the second and fourth layer. In Table C.2, $\mathbf{W}_b^{<1>}$, $\mathbf{W}_b^{<2>}$, $\mathbf{W}_b^{<3>}$ and $\mathbf{W}^{<4>}$ are matrices of the network weights. The formulation here is generic – functions $f(.)$ and $g(.)$ can assume logistic function, tan hyperbolic function or radial basis function. If we assume that the actual output is designated by $T_{(Nd,ny)}$, the error matrix between the model and the actual data is explained in matlab suitable form as:

$$V = sum((T - Y)^2, 2) \tag{C.14}$$

Calculation of Model Confidence Interval for AANN

The AANN model can be trained by derivative based or global optimization algorithms. For AANN, developing model confidence intervals applying the methods

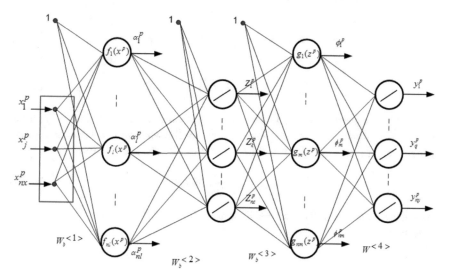

Fig. C.4 Structure of auto-associative neural network

Table C.2 Equations for AANN layer outputs

Layer	Output equation	Activation function
2	$A = f(X_b W_b^{<1>})$; $X_b = \left[\; ones(N_d, 1) \; X \; \right]$	Hyperbolic tangent
3	$Z = A_b W_b^{<2>}$; $A_b = \left[\; ones(N_d, 1) \; A \; \right]$	Linear
4	$\Phi = g(Z_b W_b^{<3>})$; $Z_b = \left[\; ones(N_d, 1) \; Z \; \right]$	Hyperbolic tangent
5	$Y = \Phi W^{<4>}$	Linear

discussed in Chap. 5 is complicated by the number of layers involved and the model having multiple outputs. Hence, in this section we follow a simplified approach.

At the first stage, the Mean of the Sum of Squared Errors (MSSR) and standard deviation are calculated. For the whole model,

$$MSSR_{total} = \frac{1}{N_d} \sum_{j=1}^{nx} \sum_{p=1}^{N_d} \left(t_j^p - y_j^p \right)^2 \tag{C.15}$$

$$\sigma_{total} = \sqrt{MSSR_{total}} \tag{C.16}$$

For an individual sensor:

$$MSSR_j = \frac{1}{N_d} \sum_{p=1}^{N_d} \left(t_j^p - y_j^p\right)^2 \tag{C.17}$$

$$\sigma_j = \sqrt{MSSR_j} \tag{C.18}$$

where, t_j^p is the actual data, y_j^p is the model output and N_d is the data size. Applying 3-sigma technique, which corresponds to 99% confidence level, the upper and lower, respectively, of the control limits for fault free operation takes the form:

$$CI_{upper,lower} = MSSR \pm \sigma \tag{C.19}$$

In fault detection, the calculated residuals are compared with CI for normal operating condition.

C.3   Model Performance Measures

Model evaluation is a requirement to guarantee that the model is indeed accurate enough to represent the system. In fault detection and diagnosis, the identified model might be a representation of the actual input output relation or the relationship formulated to capture a simulated result for a given input. In both cases, the model needs to be thoroughly validated and its adequacy tested against alternative models. To this end, there are several model performance measures that can be implemented to guarantee that the model is indeed an accurate model. Some of them are listed in Tables C.3 and C.4. In the tabulated equation, y_p^* is the measured output, $\hat{y}_p$ is the predicted output, and N_d is the number of data points. In each case, except R^2-statistics and VAF, minimum calculated value of a performance index indicates the most accurate model. For the latter two, a value close to 1 indicates an accurate model. According to Tables C.3 and C.4, all are functions of Predicted Error Sum of Squares (PRESS) except the equation for Mean Absolute Percent Error (MAP), and Average Absolute Errors (AAE). In PPE, R^2-statistics and VAF, expectation of the measured data is used to calculate for the measured data variance. In the equations for adjusted R^2-statistics, FPE and AIC, the variable n_θ stands for the umber of estimated model parameters.

Table C.3 Model performance indices

Model performance index	Equation
Predicted error sum of squares	$PRESS = \sum_{p=1}^{N_d} (y_p^* - \hat{y}_p)^2$
Mean squared error	$MSE = PRESS/N_d$
Percent mean absolute percent error	$PMAP = \frac{100}{N_d} \sum_{p=1}^{N_d} \left\| \frac{y_p^* - \hat{y}_p}{y_p^*} \right\|$
Root mean squared error	$RMSE = \sqrt{MSE}$

Table C.4 Model performance indices (cont.)

Model Performance Index	Equation
Percentage mean squared error	$PMSE = \left(RMSE \left/ \frac{1}{N_d} \sum_{p=1}^{N_d} y_p^2 \right. \right) \times 100$
Percent prediction error [4]	$PPE = \left(PRESS \left/ \sum_{p=1}^{N_d} (y_p^* - \bar{y})^2 \right. \right) \times 100$
Average absolute errors	$AAE = \frac{1}{N_d} \sum_{p=1}^{N_d} \left\| \hat{y}_p - y_p^* \right\|$
R^2-statistics	$R^2 = 1 - \left(\sum \left(y_p^* - \hat{y}_p \right)^2 \left/ \sum \left(y_p^* - \bar{y}_p \right)^2 \right. \right)$
Adjusted R^2-statistics	$R_{adj}^2 = 1 - \dfrac{\sum \left(y_p^* - \hat{y}_p \right)^2 / (N_d - n_\theta - 1)}{\sum \left(y_p^* - \bar{y}_p \right)^2 / / (N_d - 1)}$
Akaike's final prediction error	$FPE = \left(\frac{N_d + n_\theta}{N_d - n_\theta} \right) MSE$
Akaike's information criterion	$AIC = \log(MSE) + \frac{2n_\theta}{N_d}$
Minimum description length	$MDC = N_d \log(MSE) + n_\theta \log(N_d)$
Variance accounted for	$VAF = \left(1 - \frac{var(y - \hat{y})}{var(y)} \right) \times 100$

References

1. Chui, C. K. (1992). *An introduction to wavelets*. San Diego: Academic Press Limited.
2. Jackson, J. E. (1991). *A user's guide to principal components* (Vol. 587). New York: Wiley.
3. Kramer, M. A. (1991). Nonlinear principal componenet analysis using autoassociative neural networks. *AIChE Journal, 37*, 233–243.
4. Srinivasarao, M., Patwardhan, S. C., & Gudi, R. D. (2006). From data to nonlinear predictive control. 1. Identification of multivariable nonlinear state observers. *Industrial and Engineering Chemistry Research*.

Appendix D
Flowcharts Related to the Developed FDD System

D.1 Flow Charts for Data Processing and Model Training

See Figs. D.1, D.2, D.3 and D.4.

D.2 Flow Charts for Design Point and Off-Design Point Calculation

See Figs. D.5, D.6, D.7 and D.8.

D.3 Optimization Algorithms

See Figs. D.9 and D.10.

© Springer International Publishing AG 2018

T.A. Lemma, *A Hybrid Approach for Power Plant Fault Diagnostics*,
Studies in Computational Intelligence 743,
https://doi.org/10.1007/978-3-319-71871-2

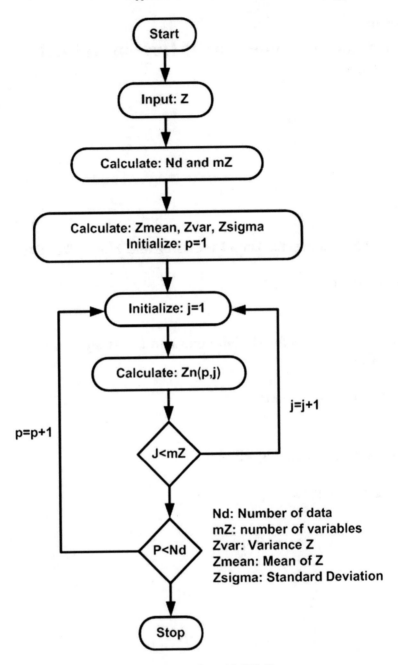

Fig. D.1 Flow chart to normalize input–output data with $N(0, 1)$

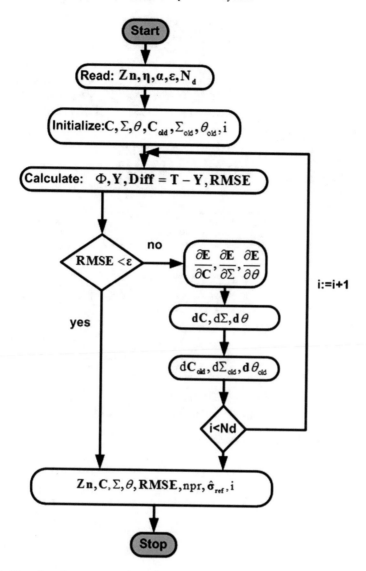

Fig. D.2 Flow chart for model training in FDEM

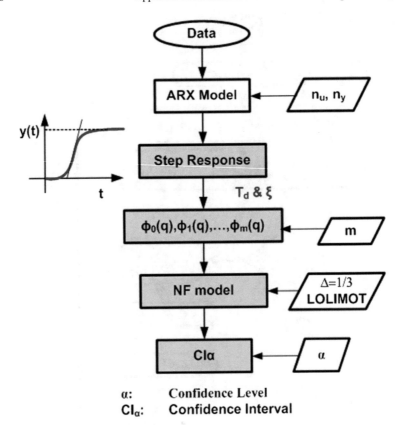

α: Confidence Level
CI_α: Confidence Interval

Fig. D.3 Flow chart for NF-OBF model training

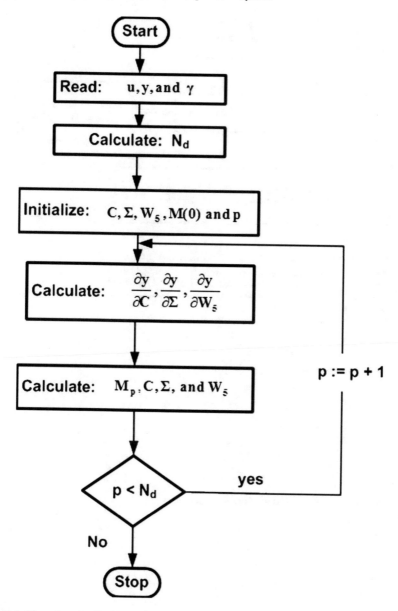

Fig. D.4 Flow chart for OBE based nonlinear model training

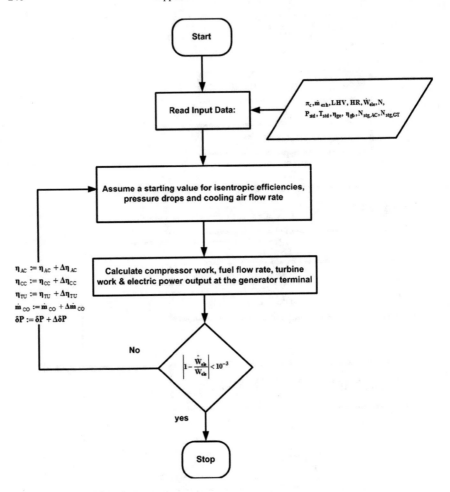

Fig. D.5 Flow chart for GTG design point calculation

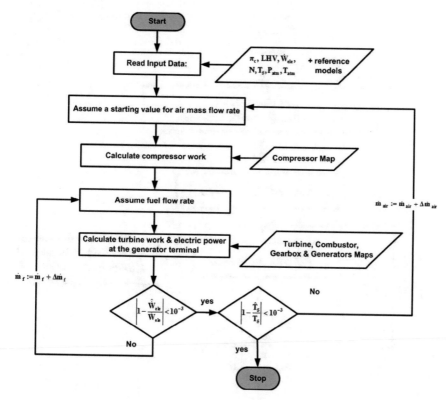

Fig. D.6 Flow chart for GTG off-design calculation

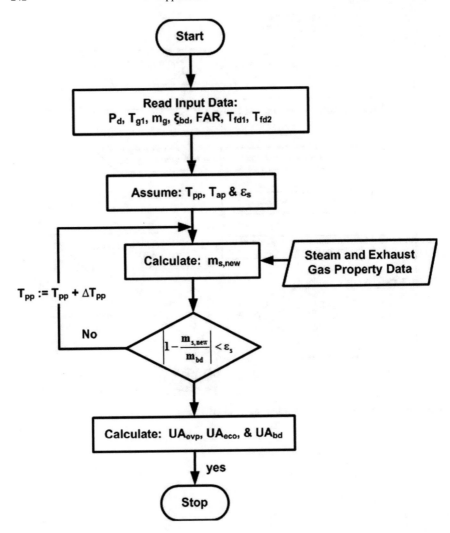

Fig. D.7 Flow chart for HRSG design point calculation

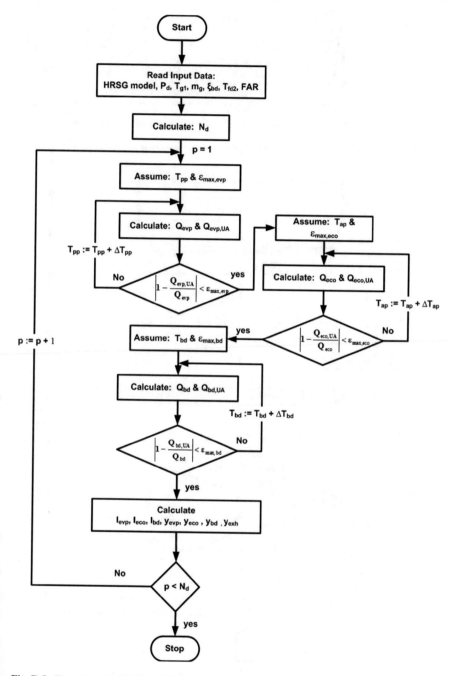

Fig. D.8 Flow chart for HRSG off-design point calculation

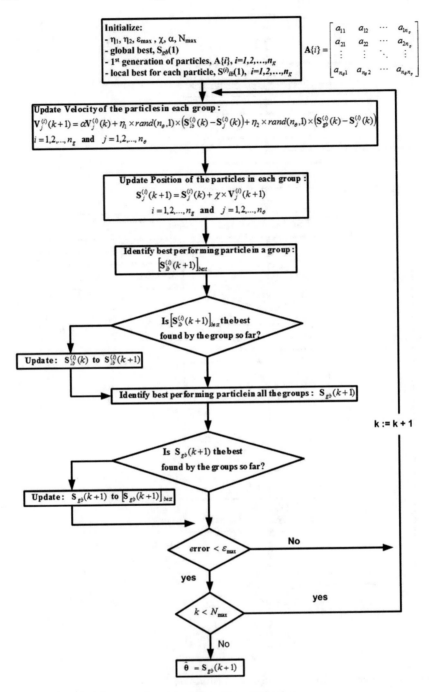

Fig. D.9 Flow chart for group based particle swarm optimization

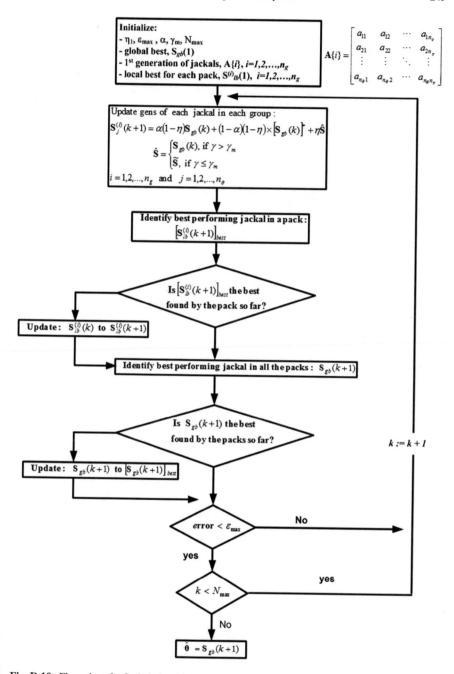

Fig. D.10 Flow chart for Jackal algorithm

Appendix E
Semi-emperical Models for GTG and HRSG

E.1 Compressor

For a given inlet condition ($\dot{m}_{air}$, T_1, P_1) and pressure ratio π_c, the properties at the outlet of the compressor are calculated by:

$$T_2 = T_1 \left\{ 1 + \frac{1}{\eta_c} (\pi_c^{(\gamma_c-1)/\gamma_c} - 1) \right\} \tag{E.1}$$

$$\dot{W}_c = \dot{m}_{air} (h_2 - h_1) \tag{E.2}$$

$$\dot{I}_c = \dot{W}_c - \dot{m}_{air} (\varepsilon_2 - \varepsilon_1) \tag{E.3}$$

$$y_c = \frac{\dot{I}_c}{\dot{I}_{c,total}} \tag{E.4}$$

where, $h_i = \int_{T_{ref}}^{T_i} C_p(T)dT$, $for\ i = 1, 2$ is enthalpy of the compressed air; γ_c and η_c are ratio of specific heats and isentropic efficiency, respectively. The correlation for the calculation of specific heat as a function of temperature is given in Appendix G. The pressure ratio and efficiency, respectively, are often read from specific compressor maps usually described as $\pi_c = f_\pi(\dot{m}_{air}, N_{sh})$ and $\eta_c = f_\eta(PR_c, N)$. Where, N_{sh} is the compressor shaft speed. For a variable geometry compressor, the maps are either based on generalized stage performance maps or relying on a fixed geometry map corrected for VIGV positions. $\dot{W}_c$, $\dot{I}_c$, $\varepsilon_i (i = 1, 2)$ and y_c stands for compressor work, exergy destruction rate, specific exergy and exergy destruction ratio, respectively. The specific exergy ε_i at the inlet or outlet of the compressor is calculated by (E.5). In fact this equation is applicable to the other systems as well.

© Springer International Publishing AG 2018
T.A. Lemma, *A Hybrid Approach for Power Plant Fault Diagnostics*,
Studies in Computational Intelligence 743,
https://doi.org/10.1007/978-3-319-71871-2

$$\varepsilon_i = C_p(T_i) \left\{ (T_i - T_{ref}) + T_{ref} \left[-\ln\left(\frac{T_i}{T_{ref}}\right) + \frac{\gamma_i - 1}{\gamma_i} \ln\left(\frac{P_i}{P_{ref}}\right) \right] \right\} \quad (E.5)$$

where, T_{ref} and P_{ref} are reference temperature and pressure, respectively. The temperature and pressure at the reference point are assumed to be 25 °C and 101.325 kPa, respectively.

E.2   Combustion Chamber

The compressed air from the compressor is combusted in the combustion chamber before it expands through the turbine. The temperature at the outlet of the combustion chamber (T_3) can be estimated from energy balance around the combustor. While the combustion reaction equation is described by the general equation

$$C_x H_y O_z + \lambda^* \left(x + \frac{y}{2} - z\right)(O_2 + 3.773 N_2) \rightarrow a CO_2 + b H_2 O + c O_2 + d CO + e N_2 \quad (E.6)$$

The energy balance is given by

$$\left(\sum_{i=1}^{n_r} n_i h_i\right)_{reac} + \eta_{cc}.FAR.LHV = (1 + FAR)\left(\sum_{j=1}^{n_p} n_j h_j\right)_{prod} \quad (E.7)$$

where, $n_i (i = 1, 2, \ldots, n_r)$ and $n_j (j = 1, 2, \ldots, n_p)$ are mole fractions of the species in the reactants and products, respectively; FAR and LHV are fuel-to-air ratio and lower heating value, respectively. The pressure loss in the combustor due to the chamber resisting air flow, high level of turbulence required for combustion and heat addition is modeled as:

$$P_3 = P_2(1 - \Delta P_{23}) \quad (E.8)$$

For the combustor, two parameters need to be calculated prior to estimating the exhaust gas temperature. The first parameter is the combustor efficiency. Adapting the procedure mentioned in Walsh and Fletcher [1], the efficiency could be read from Fig. E.1 after calculating the combustor loading that is given by (E.9).

$$\lambda_{cc} = \frac{\dot{m}}{V.P_3^{1.8}.10^{(0.00145 \times (T_3 - 400))}} \quad (E.9)$$

where, $\dot{m}$ is the mass flow rate; V is volume of the combustor; P_3 is the pressure at the inlet to the combustor; T_3 is the temperature at the inlet of the combustor. The volume can be determined from design point calculations. The second parameter, the pressure loss at off-design point, is calculated as

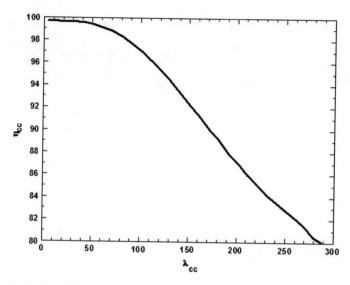

Fig. E.1 Combustion efficiency versus loading [1]

$$\frac{\Delta P R_{23'}}{P_2} = \text{PLF} \times M_{23'}^2 \gamma_2 + \left(K_1 + K_2 \left(\frac{T_3}{T_2} - 1 \right) \right) \qquad \text{(E.10)}$$

$$M_{23'} = \frac{\dot{m}_{air}}{P_2} \sqrt{\frac{R_2 T_2}{\gamma_2}} \qquad \text{(E.11)}$$

where, PLF is the pressure loss factor; K_1 and K_2 are the cold loss and hot loss, respectively. The exergy destruction and exergy destruction ration associated with the combustor can be described by the following equations.

$$\dot{I}_{cc} = \dot{m}_f \varepsilon_f - \dot{m}_c \left[(1 + \text{FAR}) \varepsilon_4 - \varepsilon_3 \right] \qquad \text{(E.12)}$$

$$y_{cc} = \frac{\dot{I}_{cc}}{\dot{I}_{total}} \qquad \text{(E.13)}$$

The fuel exergy ε_f is the LHV corrected for the grade of fuel, Table E.1.

$$\varepsilon_f = R_{NG}.\text{LHV} \qquad \text{(E.14)}$$

Table E.1 Values of energy grade function for varies forms of energy at reference temperature of 30 °C [2]

Energy form	Energy grade function (R_{NG})
Electricity	1.0
Natural gas	0.913
Steam (100 °C)	0.1385
Hot water (66 °C)	0.00921
Hot air (66 °C)	0.00596

E.3 Turbine

The hot gas temperature leaving the turbine (T_4), the power transferred to the turbine shaft $\dot{W}_t$ and exergy parameters are expressed as

$$T_4 = T_3 \left\{ 1 - \eta_t \left(1 - \left(\frac{1}{\pi_t} \right)^{(\gamma_g - 1)/\gamma_g} \right) \right\} \tag{E.15}$$

$$\dot{W}_t = \dot{m}_{exh}(h_3 - h_4) \tag{E.16}$$

$$\dot{I}_t = \dot{m}_t (\varepsilon_3 - \varepsilon_7) - \dot{W}_t \tag{E.17}$$

$$y_t = \frac{\dot{I}_t}{\dot{I}_{total}} \tag{E.18}$$

where, $\dot{m}_{exh} = \dot{m}_{air} + \dot{m}_f$ is the exhaust gas flow rate; $h_i = \int_{T_{ref}}^{T_i} C_{p,g}(T)dT$, for $i = 3, 4$ is enthalpy of the exhaust gas; $C_{p,g}(T)$ is specific heat at constant pressure; $\eta_t = f_\eta(\pi_t, N_{sh})$ and $\pi_t = P_3/P_4$ are efficiency and pressure ratio, respectively of the turbine. The pressure at the outlet of the turbine (P_4) is often limited to less than 105.825 kPa to avoid turbine trip. Using this value, π_t can be calculated and used in $\dot{m}_{exh} = f_m(\pi_t, N_{sh})$ to check if the mass balance is satisfied.

To realize the effect of cooling air on the overall performance of the GTG, the following turbine stage equations are included to the component models.

Stage-1:

$$\dot{m}_4 = \dot{m}_3 + \dot{m}_{bd,1} \tag{E.19a}$$

$$\dot{m}_4 h_4 = \dot{m}_3 h_3 + \dot{m}_{bd,1} h_2 \tag{E.19b}$$

Stage-2:

$$\dot{m}_5 = \dot{m}_4 + \dot{m}_{bd,2} \tag{E.20a}$$

$$\dot{m}_5 h_5 = \dot{m}_4 h_4 + \dot{m}_{bd,2} h_2 \tag{E.20b}$$

Stage-3:

$$\dot{m}_7 = \dot{m}_5 + \dot{m}_{bd,3} \tag{E.21a}$$

$$\dot{m}_7 h_7 = \dot{m}_5 h_5 + \dot{m}_{bd,3} h_2 \tag{E.21b}$$

E.4 Scaling Method

There are three available performance calculation techniques: scaling method, stage-stacking method and blade element method. Among the three, scaling method is the easiest. The main idea in the scaling method is that performance map of a compressor can be generated if the design conditions are known $\dot{m}_d$, π_d and η_d. The basic equations for the scaling method are:

$$\left(\frac{\pi - 1}{\pi_d - 1}\right) = \left(\frac{\pi - 1}{\pi_d - 1}\right)_{map} \tag{E.22a}$$

$$\frac{\dot{m}}{\dot{m}_d} = \left(\frac{\dot{m}}{\dot{m}_d}\right)_{map} \tag{E.22b}$$

$$\frac{\eta_{is}}{\eta_{is,d}} = \left(\frac{\eta_{is}}{\eta_{is,d}}\right)_{map} \tag{E.22c}$$

Where, the subscript "*map*" refers to the reference map. Scaling method, while it is simple to apply, it neglects compressibility effect. Besides, accuracy of the method very much depends on how close the reference map is to the actual map of the machine.

E.5 Cooling Air System Considerations

One of the auxiliary systems in a gas turbine is the air system. The purpose of the air system is to provide cooling air to the hot sections, bearing chamber, and controller circuits of the gas turbine. The air required for cooling and other purposes are extracted at a certain location in the compressor with the condition that the pressure is sufficient enough to overcome the losses in the flow path and the static pressure at the sink point is higher than the surrounding static pressure. The amount of air extracted from the compressor affects the performance of the system. For this reason, some assumptions need to be made at the design point calculations. Referring to the values given in [1], for turbine disc cooling and rim sealing a quantity of 0.5% per disc face is recommended. For bearing chamber sealing, around 0.02 kg/s per chamber is recommended. In addition to the two flows, a separate flow is required for cooling first stage stators. The amount for cooling the first stage stator vanes and rotor blades is technology dependent. Hence, for a given stator exit temperature, the percent of cooling air required may be read from the chart provided in Walsh and Fletcher [1]. After the design calculation is performed, the flows for off-design operations are calculated assuming that the cooling air flows at the deign point are choked.

$$\frac{\dot{m}_{off}\sqrt{T_{in}}}{P_{in}} = \left(\frac{\dot{m}_{chk}\sqrt{T_{in}}}{P_{in}}\right)_d \left(\left(\frac{k}{R}\right)^{\frac{1}{2}}\left(\frac{2}{k+1}\right)^{\frac{k+1}{2(k-1)}}\right)_d^{-1} \left(\left(\frac{k}{R}\right)^{\frac{1}{2}}\left(\frac{2}{k+1}\right)^{\frac{k+1}{2(k-1)}}\right)$$

(E.23)

where $\dot{m}_{chk}$ is the choked flow rate assumed at the design point, $\dot{m}_{off}$ is the air flow at the off-design point, T_{in} is the temperature at the inlet, P_{in} is the pressure at the inlet and the subscript d stands for design point.

E.6   Ducts-Off Design Performance

The gas turbine is also featured by ducts at different locations. Depending on the location and shape of the duct, there is a possibility of pressure loss. To take this effect into account while performing off-design calculations, the following equation is adopted.

$$\frac{P_{0,in} - P_{s,in}}{P_{0,in}} = \left(\frac{P_{0,in} - P_{s,in}}{P_{0,in}}\right)_d \left(\frac{\dot{m}\sqrt{T_{0,in}}}{P_{0,in}}\right)_d^{-2} \left(\frac{\dot{m}\sqrt{T_{0,in}}}{P_{0,in}}\right)^2$$

(E.24)

E.7   Generator and Gearbox

The following is assumed for the electric power at the generator terminal:

$$W_{ele} = \left(\frac{1}{\eta_{gb}\eta_{ele}}\right)\left(\dot{W}_t - \frac{\dot{W}_c}{\eta_m}\right)$$

(E.25)

Here, η_m and η_{ele} are mechanical efficiency and electrical efficiency, respectively. At part load, the efficiencies are calculated according to the empirical models given in [3].

For the gearbox:

$$\eta_{gb} = \frac{\left(\frac{\dot{W}_{out}}{\dot{W}_{in}}\right)\eta_{gb,d}}{\left(\frac{\dot{W}_{out}}{\dot{W}_{in}}\right)\eta_{gb,d} + \left(1 - \eta_{gb,d}\right)}$$

(E.26)

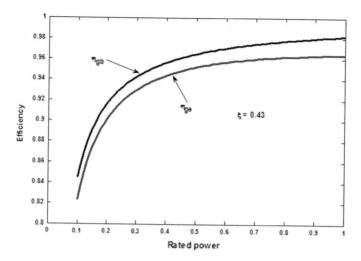

Fig. E.2 Curves of gearbox and generator efficiencies [3]

For the generator:

$$\eta_{ge} = \frac{\left(\frac{\dot{W}_{out}}{\dot{W}_{in}}\right) \eta_{ge,d}}{\left(\frac{\dot{W}_{out}}{\dot{W}_{in}}\right) \eta_{ge,d} + c_{12}\left(1 - \eta_{gb,d}\right)} \tag{E.27}$$

where,

$$c_{12} = \left\{\left(1 - \xi^*\right) + \xi^*\left(\frac{\dot{W}_{out}}{\dot{W}_{in}}\right)^2\right\}$$

The value of ξ^* is taken as 0.43. Walsh and Fletcher [1] also suggested a similar correlation. Plots of the efficiencies governed by (E.26) and (E.34) are given in Fig. E.2.

E.8 Emission Model

The prediction of green house gas emissions is as important as the performance analysis itself. The emission of NO_x is a function of combustion temperature, pressure, humidity, fuel-air ratio, and combustor geometry. There are different models available for predicting NO_x emission at part load operation. In this book, the empirical equation as stated in [4] and tested on different gas fired gas turbines is adopted.

$$NO_x = 18.1 P_3^{1.42} m_{cc,in}^{0.3} \text{FAR}^{0.72} \tag{E.28}$$

An alternative equation is also available. In this case, the emission of NO_x is given by

$$NO_x = 62 P_3^{0.5} \text{FAR}^{1.4} \exp\left(-\frac{635}{T_3}\right) \tag{E.29}$$

In (E.28) and (E.29), the unit of pressure is atm while the mass flow rate is in kg/s. The combustion temperature is measured in °C. The predictions quantify NO_x emission in ppmv at 15 °C dry air.

E.9 Performance Parameters

In the first step of missing data estimation, the component model equations are solved iteratively to determine estimated design point parameters. At each state point variation of specific heats with temperatures are taken into consideration. Often, Heat Rate (HR), Lower Heating Value (LHV) and exhaust flow rate ($\dot{m}_{exh}$) are provided. In that case, the following equations are applied to obtain the fuel flow rate and air mass flow rate, respectively.

$$\dot{m}_f = \frac{W_{ele} \times \text{HR}}{\text{LHV}} \tag{E.30}$$

$$\dot{m}_{air} = \dot{m}_{exh} - \dot{m}_f \tag{E.31}$$

In some cases, Specific Fuel Consumption (SFC) and thermal efficiency (η_{th}) may also be given. The corresponding equations are:

$$\text{SFC} = \frac{3600 \times \dot{m}_f}{\dot{W}_{ele}} \tag{E.32}$$

$$\eta_{th} = \frac{\dot{W}_{ele}}{\dot{m}_f \times \text{LHV}} \times 100 \tag{E.33}$$

For comparison purpose, the exergetic efficiency ε_{GTG} may also be included. It is given by

$$\varepsilon_{GTG} = \frac{\dot{W}_{ele}}{R_{NG} \times \dot{m}_f \times \text{LHV}} \times 100 \tag{E.34}$$

Once the component design parameters are determined, scaling method is applied to generate performance maps for the turbine stages. Off-design analysis is just solving the same set of equations for a given input. The algorithms formulated for design point and off-design point calculation are shown in Appendix D, Figs. D.5 and D.6.

E.10   Heat Recovery Steam Generator (HRSG)

The off-design performance calculation of the HRSG also requires design point parameters that include overall heat transfer coefficients, drum size, design flow rates and so on. For a commercially available HRSG, the design parameters are partially known. In the present work, the available information is combined with careful design assumptions to infer on the missing design data.

For the HRSG considered in this thesis, the design steam flow capacity is about 12 Ton/h; temperature of the hot gas at inlet to the diverter damper is ∼540 °C; exhaust gas flow rate is about 71826 kg/h. From the available Operation and Maintenance (OM) manual, inside diameters of the steam drum and mad drum, are 1042 and 860 mm, respectively. The evaporator is designed for a working pressure of about 1176.8 kPa. The output from the evaporator is a saturated steam at the design pressure. The HRSG runs by two control systems. The first controller is to control the steam pressure from the steam drum or from the evaporator. The controller manipulates the diverter damper position to vary the quantity of hot gas going to the HRSG that has an effect on the steam pressure. The second controller is referred as three-element controller. It is used to keep the level of water inside the steam drum at the set point. Signals from three sources – water level transducer, steam flow transducer, and feed water flow transducer – are used to decide on the amount of feed water going to the steam drum.

The factors critical to the successful operation of HRSG are: pinch point temperature (T_{pp}), approach temperature (T_{ap}) and pressure drop in the HRSG. The pinch point is the temperature difference between the exhaust gas temperature leaving the evaporator and the evaporator temperature. Minimum pinch seems providing better heat extraction from the exhaust gas. Nevertheless, low pinch requires high surface area of heat exchange and hence high pressure loss that is not advisable. In general, the pinch point is assumed to lie in the range of 8–22 °C [5]. Efficiency of the gas turbine is affected by the pressure drop in the system downstream of the gas turbine. In practice this pressure drop is limited to a value in the range of 200–350 mmWC.

The approach temperature – the difference between the evaporator temperature and temperature of the feed water at the inlet of the evaporator – is also need to be selected in such a way that steaming of the feed water prior to entering the evaporator is avoided. In design, it is a common practice to assume a value in the range of 5.5–11 °C [5]. The schematic diagram for the HRSG and the corresponding temperature profiles are shown in Figs. E.3 and E.4, respectively.

Once suitable values for pinch and approach temperatures are assumed, the temperature of the exhaust gas leaving the evaporator section and the temperature of the feed water entering the evaporator are calculated by (E.35) and (E.36), respectively.

$$T_{g2} = T_{evp} + \Delta T_{pp} \tag{E.35}$$

$$T_{fd3} = T_{evp} - \Delta T_{ap} \tag{E.36}$$

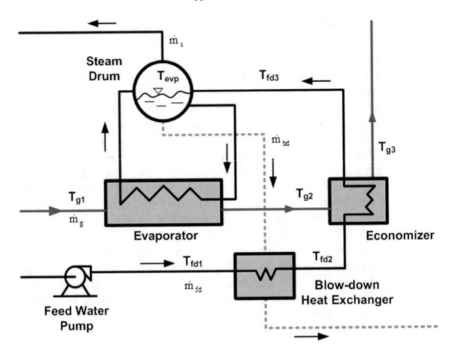

Fig. E.3 Simplified model for a heat recovery steam generator

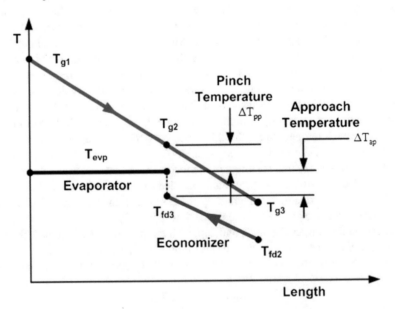

Fig. E.4 Approximate temperature profile in the HRSG

Continuous blow-down from the steam drum and intermittent blow-down from the mad drum is a common procedure to safeguard the boiler from scale formation. Assuming a continuous blow-down of quantity $\dot{m}_{bd}$, the rate of steam generation $\dot{m}_s$ in the evaporator is calculated as

$$\dot{m}_s = \frac{\dot{m}_g(h_{g1} - h_{g2})}{h_{fg} + \left(\frac{1}{1-\xi_{bd}}\right)(h_f - h_{fd3})} \tag{E.37}$$

where, ξ_{bd} is the blow-down fraction; $h_i = \int_{T_{ref}}^{T} C_{pg}(T)dT$, $(i = g1, g2, g3)$ is enthalpy of the exhaust gas; $\dot{m}_g$ is flow rate of the exhaust gas. The equations for $C_{pg}(T)$ and steam are given in Appendix G. From the design of the feed water pipe line, specifically from specification of the feed water pump, the temperature at the outlet of the pump is limited to 100 °C. In the process flow diagram corresponding to the HRSG, it is also mentioned that the feed water temperature from the well tank is considered as ~80 °C. Writing energy balance around the economizer, the temperature of exhaust gas leaving the economizer can be iteratively calculated from (E.38).

$$h_{g3} = h_{g2} - \left(\frac{\dot{m}_{fd}}{\dot{m}_g}\right)\left(h_{fd3} - h_{fd2}\right) \tag{E.38}$$

Applying energy equation for the heat exchanger between the blow-down water and the feed water from the feed water pump,

$$\dot{m}_{fd}(h_{fd2} - h_{fd1}) = \dot{m}_{bd}(h_f - h_{bd}) \tag{E.39}$$

In (E.38) and (E.39), $\dot{m}_{fd}$ and $\dot{m}_{bd}$ represent mass flow rates of the feed water and blow-down water, respectively. Once the temperatures at the inlet and outlet of the HRSG components are determined, the overall heat transfer coefficients as estimated relying on the temperatures and component duty. For evaporator,

$$UA_{evp} = \frac{\dot{Q}_{evp}}{LMTD_{evp}} \tag{E.40a}$$

$$LMTD_{evp} = \frac{T_{g1} - T_{g2}}{\ln\left(\frac{T_{g1}-T_{evp}}{T_{g2}-T_{evp}}\right)} \tag{E.40b}$$

If $\dot{Q}_{evp}$ is calculated from energy balance on the hot gas, then

$$\dot{Q}_{evp} = \dot{m}_g(h_{g1} - h_{g2}) \tag{E.41a}$$

$$UA_{evp} = \dot{m}_g C_{pg} \ln\left(\frac{T_{g1} - T_{evp}}{T_{g2} - T_{evp}}\right) \tag{E.41b}$$

For the economizer,

$$UA_{eco} = \frac{\dot{Q}_{eco}}{LMTD_{eco}} \tag{E.42a}$$

$$LMTD_{eco} = \frac{(T_{g2} - T_{fd3}) - (T_{g3} - T_{fd2})}{\ln\left(\frac{T_{g2} - T_{fd3}}{T_{g3} - T_{fd2}}\right)} \tag{E.42b}$$

Estimating the economizer duty from the heat balance in the hot gas, we have

$$\dot{Q}_{eco} = \dot{m}_g(h_{g2} - h_{g3}) \tag{E.43}$$

Following a similar approach, the overall heat transfer coefficient for the blow-down heat exchanger is

$$UA_{bd} = \frac{Q_{bd}}{LMTD_{bd}} \tag{E.44a}$$

$$LMTD_{bd} = \frac{(T_{evp} - T_{fd1}) - (T_{bd} - T_{fd2})}{\ln\left(\frac{T_{evp} - T_{fd1}}{T_{bd} - T_{fd2}}\right)} \tag{E.44b}$$

Duty of the blow-down heat exchanger can also be calculated as

$$\dot{Q}_{bd} = \dot{m}_{fd}(h_{fd2} - h_{fd1}) \tag{E.45}$$

E.11   Off-Design Calculation in the HRSG

The HRSG runs at off-design conditions depending upon the load requirement. The overall heat transfer coefficients for part load operation will be different from the design point values as they rely on the fluid flow rates and temperatures. In the sequel, the methods we have adopted in the thesis are discussed. For either the evaporator or economizer, the amount of heat exchange can be written as

$$\dot{Q} = UA.LMTD \tag{E.46}$$

$$\dot{Q} = \dot{m}_g \Delta h_g = \dot{m}_s \Delta h_s \tag{E.47}$$

At the design point, (E.46) becomes

$$\dot{Q}_d = (UA)_d LMTD_d \tag{E.48}$$

Dividing (E.46) by (E.48) results in

$$\frac{\dot{Q}}{\dot{Q}_d} = \frac{(UA)}{(UA)_d} \cdot \frac{LMTD}{LMTD_d} \tag{E.49}$$

The overall heat transfer coefficient can be approximated by

$$U = \frac{1}{1/\alpha_g + R_{f1} + R_w + R_{f2} + 1/\alpha_s} \tag{E.50}$$

where, α_g and α_s are convective heat transfer coefficients on the hot gas and steam side, respectively; R_{f1}, R_{f2} and R_w are heat transfer coefficients for conduction; While R_{f1} and R_{f2} due to fouling, R_w is related to heat exchanger shell thickness. Due to their minimal contribution, the coefficients for heat conduction can be neglected. For the evaporator and economizer, α_s is about 10 to 100 times bigger than α_g. In light of these, UA value for part load operation is approximated by

$$UA = (UA)_d \frac{\alpha_g}{\alpha_{g,d}} \tag{E.51}$$

The convective heat transfer coefficient on the gas side can be calculated from the equation for Nusselt number Nu_g. That is,

$$\alpha_g = \frac{k_g Nu_g}{d} = \frac{k_g}{d} C.Re^m \stackrel{n}{Pr} \tag{E.52}$$

where, k_g is thermal conductivity; C, m and n are constants that rely on geometry of the heat transfer surface. The Prandtl number Pr does not vary much with temperature. The Reynolds number, however, is affected by temperature since it is a function of density ρ_g and dynamic viscosity μ_g of the fluid. The Reynolds number is defined as

$$Re = \frac{V_g \rho_g d}{\mu_g}$$

Now, using the mass flow rate in the Reynolds number, we get

$$Re = \left(\frac{d}{A}\right)\left(\frac{\dot{m}_g}{\mu_g}\right) \tag{E.53}$$

Using (E.53) in (E.52) yields,

$$\alpha_g = \left(\frac{C}{A}\right) d^{m-1}.k_g \left(\frac{\dot{m}_g}{\mu_g}\right)^m \stackrel{n}{Pr} \tag{E.54}$$

The use of (E.54) in (E.51) eliminates the constants and geometric parameters. Hence,

$$\frac{UA}{(UA)_d} = \left(\frac{k_g}{k_{g,d}}\right)\left(\frac{\mu_{g,d}}{\mu_g}\right)^m \left(\frac{\dot{m}_g}{\dot{m}_{g,d}}\right)^m \tag{E.55}$$

Table E.2 Exergetic parameters for the HRSG

Component	Exergetic parameters
Evaporator	$\dot{I}_{evp} = \dot{m}_g C_{p,g} T_{std} \ln\left(\frac{T_{g1}}{T_{g2}}\right); \ y_{evp} = \frac{\dot{I}_{evp}}{\dot{I}_{total}}$
Economizer	$\dot{I}_{eco} = \dot{m}_g C_{p,g} T_{std} \ln\left(\frac{T_{g2}}{T_{g3}}\right); \ y_{eco} = \frac{\dot{I}_{eco}}{\dot{I}_{total}}$
Blow-down heat exchanger	$\dot{I}_{bd} = \dot{m}_{bd} T_{std}(s_{bd,1} - s_{bd,2}); \ y_{bd} = \frac{\dot{I}_{bd}}{\dot{I}_{total}}$
Exhaust duct	$\dot{I}_{exh} =$ $\dot{m}_g C_{p,g}(T_{g3} - T_{atm}) - \dot{m}_g C_{p,g} T_{std} \ln\left(\frac{T_{g1}}{T_{atm}}\right);$ $y_{exh} = \frac{\dot{I}_{exh}}{\dot{I}_{total}}; \ \dot{I}_{total} = \dot{I}_{evp} + \dot{I}_{eco} + \dot{I}_{bd} + \dot{I}_{exh}$

The constant m is often assumed in the range of 0.57 to 0.65. For our case a value of 0.6 is taken for both the evaporator and economizer. Once the overall heat transfer coefficient is calculated from (E.55), the off-design point exhaust gas temperature leaving the evaporator can be estimated from

$$T_{g2} = T_{evp} + (T_{g1} - T_{evp}) \exp\left(-\frac{UA}{\dot{m}_g C_{p,g}}\right) \tag{E.56}$$

For the economizer, while (E.42a) and (E.55) are applicable, the calculation for exit temperatures needs iterative processing of the equations together with the models for the blow-down heat exchanger. In the blow-down heat exchanger, the heat transfer fluids are both hot water.

For the blow-down heat exchanger, at the design point we have assumed equal convective heat transfer coefficients on the hot and cold side. Accordingly, the overall heat transfer coefficient for the off-design point could be estimated applying the following equation:

$$UA_{bd} = 2 \times (UA_{bd,d}) \left(\frac{1}{\left(\frac{\dot{m}_{fd,d}}{\dot{m}_{fd}}\right)^{0.8} + \left(\frac{\dot{m}_{bd,d}}{\dot{m}_{bd}}\right)^{0.6}} \right) \tag{E.57}$$

E.12 Exergetic Parameters for the Evaporator, Economizer and Blow-Down Heat Exchanger

The exergy destruction and exergy destruction ratio for each component are calculated applying the exergy equation for flow streams, (E.5). Table E.2 provides the resulting equations.

References

1. Walsh, P. P., & Fletcher, P. (2004). *Gas turbine performance*. Malden: Blackwell Science Ltd.
2. Dincer, I., & Rosen, M. A. (2007). *Exergy: Energy, environment, and sustainable development*. Oxford: Elsevier Ltd.
3. Haglind, F. (2010). Variable geometry gas turbines for improving the part-load performance of marine combined cycles - gas turbine performance. *Energy, 31*, 467–476.
4. Razak, A. M. Y. (2007). *Industrial gas turbines performance and operability*. Boca Raton: Woodhead Publishing Limited and CRC Press LLC.
5. Boyce, M. P. (2006). *Gas Turbine Engineering Handbook*. Boston: Gulf Professional Publishing.

Appendix F
Schematic Diagram, Design Point or Alarm Setting Data for SAC, HRSG and GTG

F.1 Schematic Diagram for the SAC

See Fig. F.1.

F.2 Known Design Point and Alarm Related Data for the SAC, HRSG and GTG

See Tables F.1, F.2, F.3, F.4, F.5, F.6, F.7, F.8, F.9 and F.10.

© Springer International Publishing AG 2018
T.A. Lemma, *A Hybrid Approach for Power Plant Fault Diagnostics*,
Studies in Computational Intelligence 743,
https://doi.org/10.1007/978-3-319-71871-2

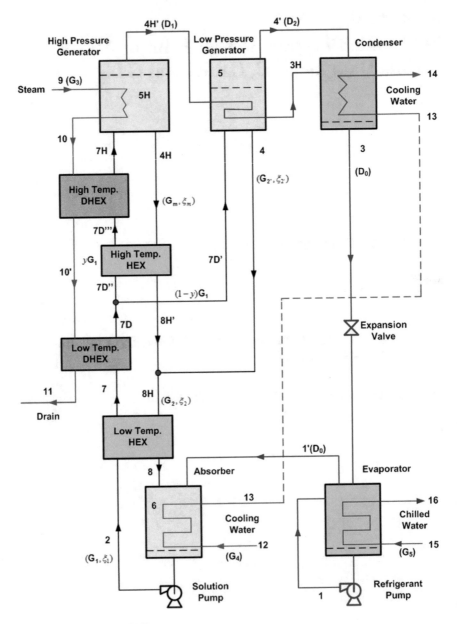

Fig. F.1 Flow sheet for SAC

Table F.1 Design point cooling water and chilled water data for the SAC

Parameter		Unit	Value
Refrigeration capacity		kW	4395
Chilled water	Inlet temperature	°C	13.5
	Outlet temperature	°C	6.0
	Flow rate	M³/h	504
	Fouling factor	M²k/w	0.000086
	Pass number	Pass	2
	Pressure drop	kPa	44
	Max. working pressure	MPa	1.37
Cooling water	Inlet temperature	°C	32
	Outlet temperature	°C	39.5
	Flow rate	M³/h	920
	Fouling factor	M²k/w	0.000086
	Pass number	Pass	2(ABS) + 1(COND)
	Pressure drop	kPa	49
	Max. working pressure	MPa	0.78

Table F.2 Design point values related to the heating medium in the SAC

Parameter		Unit	Value
Heating medium	Steam pressure	MPa	0.78
	Steam temperature	°C	Saturated temperature
	Steam consumption	kg/h	5500
	Drain temperature	°C	90
	Drain pressure	MPa	0.06

Table F.3 Summary of design point data for the 12 Ton/h HRSG

Parameter	Units	Value
Hot gas HRSG inlet temperature, T_{g1}	°C	540
Hot gas Economizer inlet temperature, T_{g2}	°C	209
Economizer inlet temperature, T_{fd2}	°C	93
Economizer outlet temperature, T_{fd3}	°C	176
Feed water valve position	%	100
Hot gas Bypass stack temperature, T_{g4}	°C	540
Diverter damper position	%	100
Drum pressure	kPa	1176.9
Drum water level	%	60
Steam flow rate	kg/s	3.3333
Feed water flow rate	kg/s	3.5461
Steam line pressure	kPa	1176.9

Table F.4 Design point values of the 5.2 MW GTG parameters as obtained from manufacturer supplied data

Parameter		Units	Value
Electric power		kW	4917
Lower heating value		kJ/kg.K	47939
Fuel flow rate		kg/s	0.3167
Compressor	Number of stages	–	12
	Compressor discharge pressure	kPa	1176.9
	Pressure ratio	–	11.615
	Number of VSVs	–	3
	Air flow rate	kg/s	21.0324
Turbine	Number of stages	–	3
	Temperature T_5	°C	676
	Exhaust temperature	°C	482
	Exhaust flow	kg/s	21.349
Generator efficiency		1	0.964
Gearbox efficiency		1	0.982

Table F.5 Design point data for the 5.2 MW GTG parameters as calculated by the design point algorithm

Parameter		Units	Value
Compressor efficiency		1	0.8551
Combustor efficiency		1	0.98
Combustor volume		m^3	0.3468
Turbine stage efficiency		1	0.899
Turbine mass flow	Stage 1	kg/s	20.38
	Stage 2	kg/s	20.49
	Stage 3	kg/s	20.59
Turbine stage pressure ratio		1	2.21
Cooling air	NGV	kg/s	0.3155
	Disc cooling per stage	kg/s	0.1052
Air for control systems		kg/s	0.1052
Air for bearing chamber		kg/s	0.5468
Inlet duct pressure loss		%	0.5
Compressor exit diffuser loss		%	2
Pressure loss in the combustion chamber		%	2

Table F.6 Set of measurable process variables of the gas turbine generator: part-I

Symbol		Units	Process variable
Gas path	T_1	°C	Air temperature at the inlet to the compressor
	θ_{IGV}	%	VIGV position
	$\dot{m}_f$	kg/h	Fuel flow rate
	P_2	kPa	Compressor discharge pressure
	$\dot{W}_{ele}$	kW	Electric power output
	T_5	°C	Temperature at the inlet to the 3rd stage of GT
	x_{mp}	%	Gas fuel meter position
	P_{df}	kPa	Gas fuel control differential pressure
	P_{vc}	kPa	Gas valve check
	x_{mc}	%	Gas fuel meter command
	x_{pvc}	%	Main pilot valve command
	P_{gp}	kPa	Gas pressure
	T_{enc}	°C	Temperature inside the turbine enclosure

Table F.7 Set of measurable process variables of the gas turbine generator: part-II

Symbol		Units	Process variable
Lubrication system	T_{lb1}	°C	Temperature of lube oil
	T_{lb2}	°C	Temperature of oil in the lube tank
	T_{lb3}	°C	Temperature of oil at the thrust bearing
	T_{lb4}	°C	Temperature of oil at the generator exciter bearing
	T_{lb5}	°C	Temperature of oil at the generator driver bearing
	P_{lb1}	kPa	Oil pressure in the lubricant tank
Generator winding	T_{wa}	°C	Temperature of generator winding-A
	T_{wb}	°C	Temperature of generator winding-B
	T_{wc}	°C	Temperature of generator winding-C
Turbine stage temperature sensors	T_{c1}	°C	Temperature at the inlet to the 3rd stage of GT
	T_{c2}	°C	Temperature at the inlet to the 3rd stage of GT
	T_{c3}	°C	Temperature at the inlet to the 3rd stage of GT
	T_{c4}	°C	Temperature at the inlet to the 3rd stage of GT
	T_{c5}	°C	Temperature at the inlet to the 3rd stage of GT
	T_{c6}	°C	Temperature at the inlet to the 3rd stage of GT

Table F.8 Measurable signals in the HRSG

Symbol	Unit	Process variable
T_{g1}	°C	Temperature of exhaust gas at the inlet to the HRSG
T_{g2}	°C	Temperature of exhaust gas at the inlet to the economizer
$T_{fd,2}$	°C	Temperature of feed water at the inlet to the economizer
$T_{fd,3}$	°C	Temperature of feed water at the outlet to the economizer
x_{bfd}	%	Feed water valve position
T_{g4}	°C	Temperature of exhaust gas in the bypass stack
θ_{dd}	%	Diverter damper position
P_D	kPa	Drum pressure
H_D	%	Drum water level
$\dot{m}_s$	kg/s	Flow rate of main steam
$\dot{m}_{fd}$	kg/s	Flow rate of feed water
P_s	kPa	Steam line pressure

Table F.9 Measurable signals in the SAC

Symbol	Unit	Process variable
$\dot{m}_{ch}$	m³/s	Chilled water flow rate
x_{bfd}	%	Chilled water valve percentage opening
T_{ch2}	°C	Temperature of chilled water at the outlet of the SAC
T_{ch1}	°C	Temperature of chilled water at the inlet to the SAC
P_s	kPa	Steam pressure
$\dot{m}_s$	m³/s	Steam flow rate
T_{co2}	°C	Temperature of cooling water at the outlet of the SAC
T_{co1}	°C	Temperature of cooling water at the inlet to the SAC

Table F.10 GTG alarm setting

Parameter	Units	Value
Inlet temperature T_1	°C	32
Electric power	Kw	4200
Compressor discharge pressure	kPa	1176.9
Temperature T_5	°C	676
Air inlet differential pressure	kPa	1.9
Enclosure temperature	°C	70
Fuel flow	kg/h	1550
Gas pressure	kPa	1900
Control differential pressure	kPa	83
Valve check	kPa	1900

(continued)

Table F.10 (continued)

Parameter	Units	Value
Lube oil header temperature	°C	73.5
Lube tank temperature	°C	76
Lube oil header pressure	kPa	172
Lube tank pressure	kPa	2
Lube oil level	%	50
Engine thrust bearing temperature	°C	121
Generator exciter bearing oil temperature	°C	121
Generator driver bearing oil temperature	°C	121

Appendix G
Thermodynamic Properties of Air, Combustion Products, Steam and LiBr + H$_2$O Solution

G.1 Property Correlations for a Dry Air

Specific heat at constant pressure in kJ/kg for gases

$$C_p(T) = \sum_{j=1}^{9} A_{j-1} T_z^{j-1} \tag{G.1}$$

where, $T_z = \frac{T}{1000}$ and T in °K

And, enthalpies in MJ/kg can be estimated from

$$h(T) = A_9 + \sum_{j=1}^{9} \left(\frac{1}{j}\right) A_{j-1} T_z^{j} \tag{G.2}$$

G.2 Property Correlations for Combustion Products

Specific heat of combustion gases at constant pressure and in kJ/kg can be calculated from,

$$C_p(T) = \sum_{j=1}^{9} A_{j-1} T_z^{j-1} + \left(\frac{FAR}{FAR+1}\right) \sum_{j=1}^{8} B_{j-1} T_z^{j-1} \tag{G.3}$$

where, FAR is the fuel to air ratio, $T_z = \frac{T}{1000}$ and T in °K. $B_0 = -0.718874$; $B_1 = 8.747481$; $B_2 = -15.863157$; $B_3 = 17.254096$; $B_4 = -10.233795$; $B_5 = 3.081778$; $B_6 = -0.361112$; $B_7 = -0.003919$; $B_8 = 0.0555930$; $B_9 = -0.0016079$ (Tables G.1 and G.2).

© Springer International Publishing AG 2018
T.A. Lemma, *A Hybrid Approach for Power Plant Fault Diagnostics*,
Studies in Computational Intelligence 743,
https://doi.org/10.1007/978-3-319-71871-2

Table G.1 Coefficients for (G.1) and (G.2)

	Dry air	O_2	N_2	CO_2	H_2O
A_0	0.9992313	1.006450	1.075132	0.408089	1.937043
A_1	0.236688	−1.047869	−0.252297	2.027201	−0.967916
A_2	−1.852148	3.729558	0.341859	−2.405549	3.338905
A_3	6.083152	−4.934172	0.523944	2.039166	−3.652122
A_4	−8.893933	3.284147	−0.888984	−1.163088	2.332470
A_5	7.097112	1.095203	0.442621	0.381364	−0.819451
A_6	−3.234725	0.145737	−0.074788	−0.052763	0.118783
A_7	0.794571	–	–	–	–
A_8	−0.081873	–	–	–	–
A_9	0.422178	0.369790	0.443041	0.366740	2.860773
A_{10}	0.001053	0.000491	0.0012622	0.001736	−0.000219

Table G.2 Composition of dry air

	By mole, (%)	By mass, (%)
N_2	78.06	75.52
O_2	20.95	23.14
Ar	0.93	1.28
CO_2	0.03	0.05
Ne	0.002	0.001

Specific enthalpy for combustion products of kerosene diesel fuel in dry air (MJ/kg) is

$$h(T) = A_9 + \sum_{j=1}^{9} \left(\frac{1}{j}\right) A_{j-1} T_z^j + B_8 + \left(\frac{FAR}{FAR+1}\right) \sum_{j=1}^{7} \left(\frac{1}{j}\right) B_{j-1} T_z^j \quad \text{(G.4)}$$

G.3 Property Correlations for Water

G.3.1 Enthalpies of Saturated and Superheated Steam

$$Y_S = A_0 + A_1 T_C^{1/3} + A_2 T_C^{5/6} + A_3 T_C^{7/8} + \sum_{j=1}^{7} E_j T_C^j \quad \text{(G.5)}$$

$$T_C = \left(1 - \frac{T_S}{T_{CR}}\right), \ T_{CR} = 647.3\,\text{K}$$

Table G.3 Correlation coefficients for $h_f(T_S)$

$273.16 \leq T_S < 300\,\text{K}$	$300 \leq T_S < 600\,\text{K}$	$600 \leq T_S \leq 647.3\,\text{K}$
$A_0 = 0.0$	$A_0 = 8.839230108E - 1$	$A_0 = 1.0$
$A_1 = 0.0$	$A_1 = 0.0$	$A_1 = -4.4105780E - 1$
$A_2 = 0.0$	$A_2 = 0.0$	$A_2 = -5.52255517$
$A_3 = 0.0$	$A_3 = 0.0$	$A_3 = 6.43994847$
$E_1 = 6.24698837E + 2$	$E_1 = -2.67172935$	$E_1 = 1.64578795E$
$E_2 = -2.34385369E + 3$	$E_2 = 6.22640035$	$E_2 = -1.30574143$
$273.16 \leq T_S < 300\,\text{K}$	$300 \leq T_S < 600\,\text{K}$	$600 \leq T_S \leq 647.3\,\text{K}$
$E_3 = -9.50812101E + 3$	$E_3 = -1.31789573E + 1$	$E_3 = 0.0$
$E_4 = 7.16287928E + 4$	$E_4 = -1.91322436$	$E_4 = 0.0$
$E_5 = -1.63535221E + 5$	$E_5 = 6.87937653E + 1$	$E_5 = 0.0$
$E_6 = 1.66531093E + 5$	$E_6 = -1.24819906E + 2$	$E_6 = 0.0$
$E_7 = -6.47854585E + 4$	$E_7 = 7.21435404E + 1$	$E_7 = 0.0$

Table G.4 Correlation coefficients for $h_{fg}(T_S)$ and $h_g(T_S)$

$h_{fg}(T_S),\ 273.16 \leq T_S \leq 647.3\,\text{K}$	$h_g(T_S),\ 273.16 \leq T_S \leq 647.3\,\text{K}$
$A_0 = 0.0$	$A_0 = 1$
$A_1 = 7.79221E - 1$	$A_1 = 4.57874342E - 1$
$A_2 = 4.62668$	$A_2 = 5.08441288$
$A_3 = -1.07931$	$A_3 = -1.48513244$
$E_1 = -3.87446$	$E_1 = -4.81351884$
$E_2 = 2.94553$	$E_2 = 2.69411792$

Table G.5 Correlation coefficients for $h_{fg}(T_S)$ and $h_g(T_S)$ (continued)

$h_{fg}(T_S),\ 273.16 \leq T_S \leq 647.3\,\text{K}$	$h_g(T_S),\ 273.16 \leq T_S \leq 647.3\,\text{K}$
$E_3 = -8.06395$	$E_3 = -7.39064542$
$E_4 = 1.15633E + 1$	$E_4 = 1.04961689E + 1$
$E_5 = -6.02884$	$E_5 = -5.46840036$
$E_6 = 0.0$	$E_6 = 0.0$
$E_7 = 0.0$	$E_7 = 0.0$

$$h_f(T_S) = Y_S h_{f,CR}$$

$$h_{fg}(T_S) = Y_S h_{f,CR}$$

$$h_g(T_S) = Y_S h_{f,CR}$$

where, $h_{f,CR} = 2009.3\,\text{kJ/kg}$ (Tables G.3, G.4 and G.5).

G.4   Property Correlations for LiBr-H$_2$0 Solution

Enthalpy of LiBr-H2O solution may be calculated as [1],

$$H(T, \xi) = (A_0 + A_1\xi)T + (B_0 + B_1\xi)T^2 + (D_0 + D_1\xi + D_2\xi^2 + D_3\xi^3) \quad (G.6)$$

where, ξ is solution concentration. The equation is valid for $20 \leq T \leq 210\,^{\circ}C$ and $40 \leq \xi \leq 65wt\%$. $A_0 = 3.462023$; $A_1 = -2.679895E - 2$; $B_0 = 1.3499E - 2$; $B_1 = -6.55E - 6$; $D_0 = 162.81$; $D_1 = -6.0418$; $D_2 = 4.5348E - 3$; $D_3 = 1.2053E - 3$.

Reference

1. Kaita, Y. (2001). Thermodynamic properties of Lithium Bromide-water solutions ar high temperatures. *International Journal of Refrigeration*, *24*, 374–390.

Appendix H
Fault Names and Input–Output Relations for GTG, HRSG, and SAC

See Tables H.1, H.2, H.3, H.4, H.5, H.6, H.7, H.8 and Figs. H.1, H.2 and H.3.

Table H.1 Set of sensor faults in the gas turbine generator: part-I

Fault name	Fault description
f_1	Air temperature, T_1, sensor fault
f_2	VIGV position, θ_{IGV}, sensor fault
f_3	Fuel flow rate, $\dot{m}_f$, sensor fault
f_4	Compressor discharge pressure, P_2, sensor fault
f_5	Electric power output, $\dot{W}_{ele}$, sensor fault
f_6	Temperature at the inlet to the 3rd stage of GTG, T_5, sensor fault
f_7	Gas fuel meter position, x_{mp}, sensor fault
f_8	Gas fuel control differential pressure, P_{df}, sensor fault
f_9	Gas valve check, P_{vc}, sensor fault
f_{10}	Gas fuel meter command, x_{mc}, sensor fault
f_{11}	Main pilot valve command, x_{pvc}, sensor fault
f_{12}	Gas Pressure, P_{gp}, sensor fault
f_{13}	Temperature inside the turbine enclosure, T_{enc}, sensor fault

© Springer International Publishing AG 2018
T.A. Lemma, *A Hybrid Approach for Power Plant Fault Diagnostics*,
Studies in Computational Intelligence 743,
https://doi.org/10.1007/978-3-319-71871-2

Table H.2 Set of faults in the gas turbine generator: part-II

Fault name	Fault description
f_{14}	Lube oil temperature, T_{lb1}, sensor fault
f_{15}	Oil in the lube tank temperature, T_{lb2}, sensor fault
f_{16}	Thrust bearing oil temperature, T_{lb3}, sensor fault
f_{17}	Generator exciter bearing oil temperature, T_{lb4}, sensor fault
f_{18}	Generator driver bearing oil temperature, T_{lb5}, sensor fault
f_{19}	Lube tank oil pressure, P_{lb1}, sensor fault
f_{20}	Generator winding-A temperature, T_{wa}, sensor fault
f_{21}	Generator winding-B temperature, T_{wb}, sensor fault
f_{22}	Generator winding-C temperature, T_{wc}, sensor fault
f_{23}	GT 3rd stage inlet temperature (T_{c1}) sensor fault
f_{24}	GT 3rd stage inlet temperature (T_{c2}) sensor fault
f_{25}	GT 3rd stage inlet temperature (T_{c3}) sensor fault
f_{26}	GT 3rd stage inlet temperature (T_{c4}) sensor fault
f_{27}	GT 3rd stage inlet temperature (T_{c5}) sensor fault
f_{28}	GT 3rd stage inlet temperature (T_{c6}) sensor fault
f_{29}	Suction duct differential pressure (ΔP_{SD}) sensor fault

Table H.3 Set of residual equations for the GTG and for pth data point: part-I

s_j	Fault detection test	Detail of u_j
s_1	$r_1 = \left\| T_1 - \hat{T}_1(u_1) \right\| \leq CI_{\alpha,1}$	$u_1 = [T_{enc}]$
s_2	$r_2 = \left\| \theta_{IGV} - \hat{\theta}_{IGV}(u_2) \right\| \leq CI_{\alpha,2}$	$u_2 = [P_2 \ \dot{W}_{ele} \ T_5]^T$
s_3	$r_3 = \left\| \dot{m}_f - \hat{\dot{m}}_f(u_3) \right\| \leq CI_{\alpha,3}$	$u_3 = [P_2 \ \dot{W}_{ele} \ T_5]^T$
s_4	$r_4 = \left\| P_2 - \hat{P}_2(u_4) \right\| \leq CI_{\alpha,4}$	$u_4 = [\theta_{IGV} \ \dot{m}_f]^T$
s_5	$r_5 = \left\| \dot{W}_{ele} - \hat{\dot{W}}_{ele}(u_5) \right\| \leq CI_{\alpha,5}$	$u_5 = [\theta_{IGV} \ \dot{m}_f]^T$
s_6	$r_6 = \left\| T_5 - \hat{T}_5(u_6) \right\| \leq CI_{\alpha,6}$	$u_6 = [\theta_{IGV} \ \dot{m}_f]^T$
s_7	$r_7 = \left\| x_{mp} - \hat{x}_{mp}(u_7) \right\| \leq CI_{\alpha,7}$	$u_7 = [\theta_{IGV} \ P_{df} \ \dot{W}_{ele}]^T$
s_8	$r_8 = \left\| P_{df} - \hat{P}_{df}(u_8) \right\| \leq CI_{\alpha,8}$	$u_8 = [P_2 \ P_{gp}]^T$
s_9	$r_9 = \left\| P_{vc} - \hat{P}_{vc}(u_9) \right\| \leq CI_{\alpha,9}$	$u_9 = [\dot{m}_f \ P_{df}]^T$
s_{10}	$r_{10} = \left\| x_{mc} - \hat{x}_{mc}(u_{10}) \right\| \leq CI_{\alpha,10}$	$u_{10} = [\theta_{IGV} \ P_{df} \ \dot{W}_{ele} \ T_5]^T$
s_{11}	$r_{11} = \left\| x_{pvc} - \hat{x}_{pvc}(u_{11}) \right\| \leq CI_{\alpha,11}$	$u_{11} = [\theta_{IGV} \ P_{df} \ \dot{W}_{ele} \ T_5]^T$
s_{12}	$r_{12} = \left\| P_{gp} - \hat{P}_{gp}(u_{12}) \right\| \leq CI_{\alpha,12}$	$u_{12} = [\theta_{IGV} \ \dot{m}_f]^T$
s_{13}	$r_{13} = \left\| T_{enc} - \hat{T}_{enc}(u_{13}) \right\| \leq CI_{\alpha,13}$	$u_{13} = [\theta_{IGV} \ \dot{m}_f]^T$
s_{29}	$r_{29} = \left\| \Delta \hat{P}_{SD} - \Delta \hat{P}_{SD}(u_{29}) \right\| \leq CI_{\alpha,29}$	$u_{29} = [\theta_{IGV} \ \dot{m}_f]^T$

Table H.4 Set of residual equations for the GTG and for pth data point: part-II

s_j	Fault detection test	Detail of u_j
s_{14}	$r_{14} = \left| T_{lb1} - \hat{T}_{lb1}(u_{14}) \right| \leq CI_{\alpha,14}$	$u_{14} = [T_{lb2}\ P_{lb1}]^T$
s_{15}	$r_{15} = \left| T_{lb2} - \hat{T}_{lb2}(u_{15}) \right| \leq CI_{\alpha,15}$	$u_{15} = [T_{lb3}\ T_{lb1}]^T$
s_{16}	$r_{16} = \left| T_{lb3} - \hat{T}_{lb3}(u_{16}) \right| \leq CI_{\alpha,16}$	$u_{16} = [T_{lb4}\ T_{lb2}]^T$
s_{17}	$r_{17} = \left| T_{lb4} - \hat{T}_{lb4}(u_{17}) \right| \leq CI_{\alpha,17}$	$u_{17} = [T_{lb5}\ T_{lb3}]^T$
s_{18}	$r_{18} = \left| T_{lb5} - \hat{T}_{lb5}(u_{18}) \right| \leq CI_{\alpha,18}$	$u_{18} = [P_{lb1}\ T_{lb4}]^T$
s_{19}	$r_{19} = \left| P_{lb1} - \hat{P}_{lb1}(u_{19}) \right| \leq CI_{\alpha,19}$	$u_{19} = [T_{lb1}\ T_{lb5}]^T$
s_{20}	$r_{20} = \left| T_{wa} - \hat{T}_{wa}(u_{20}) \right| \leq CI_{\alpha,20}$	$u_{20} = T_{wb}$
s_{21}	$r_{21} = \left| T_{wb} - \hat{T}_{wb}(u_{21}) \right| \leq CI_{\alpha,21}$	$u_{21} = T_{wc}$
s_{22}	$r_{22} = \left| T_{wc} - \hat{T}_{wc}(u_{22}) \right| \leq CI_{\alpha,22}$	$u_{22} = T_{wa}$
s_{23}	$r_{23} = \left| T_{c1} - \hat{T}_{c1}(u_{23}) \right| \leq CI_{\alpha,23}$	$u_{23} = [T_{c2}\ T_{c6}]^T$
s_{24}	$r_{24} = \left| T_{c2} - \hat{T}_{c2}(u_{24}) \right| \leq CI_{\alpha,24}$	$u_{24} = [T_{c1}\ T_{c3}]^T$
s_{25}	$r_{25} = \left| T_{c3} - \hat{T}_{c3}(u_{25}) \right| \leq CI_{\alpha,25}$	$u_{25} = [T_{c2}\ T_{c4}]^T$
s_{26}	$r_{26} = \left| T_{c4} - \hat{T}_{c4}(u_{26}) \right| \leq CI_{\alpha,26}$	$u_{26} = [T_{c3}\ T_{c5}]^T$
s_{27}	$r_{27} = \left| T_{c5} - \hat{T}_{c5}(u_{27}) \right| \leq CI_{\alpha,27}$	$u_{27} = [T_{c4}\ T_{c6}]^T$
s_{28}	$r_{28} = \left| T_{c6} - \hat{T}_{c6}(u_{28}) \right| \leq CI_{\alpha,28}$	$u_{28} = [T_{c1}\ T_{c5}]^T$

Table H.5 Set of sensor faults in the HRSG

Fault name	Process variable
f_1	Exhaust gas temperature, T_{g1}, sensor fault
f_2	Exhaust gas temperature, T_{g2}, sensor fault
f_3	Feed water temperature, T_{fd2}, sensor fault
f_4	Feed water temperature, T_{fd3}, sensor fault
f_5	Feed water valve position, x_{bfd}, sensor fault
f_6	Exhaust gas temperature, T_{g4}, sensor fault
f_7	Diverter damper position, θ_{dd}, sensor fault
f_8	Drum pressure, P_D, sensor fault
f_9	Drum water level, H_D, sensor fault
f_{10}	Main steam flow rate, $\dot{m}_s$, sensor fault
f_{11}	Feed water flow rate, $\dot{m}_{fd}$, sensor fault
f_{12}	Steam line pressure, P_s, sensor fault

Table H.6 Set of residual equations for the HRSG and for pth data point

s_j	Fault detection test	Detail of u_j
s_1	$r_1 = \left| T_{g1} - \hat{T}_{g1}(u_1) \right| \le CI_{\alpha,1}$	$u_1 = [T_{g2}\ T_{fd3}\ x_{fd}\ \theta_{dd}]^T$
s_2	$r_2 = \left| T_{g2} - \hat{T}_{g2}(u_2) \right| \le CI_{\alpha,2}$	$u_2 = [T_{g1}\ T_{fd3}\ \theta_{dd}]^T$
s_3	$r_3 = \left| T_{fd,2} - \hat{T}_{fd,2}(u_3) \right| \le CI_{\alpha,3}$	$u_3 = [x_{fd}\ P_D]^T$
s_4	$r_4 = \left| T_{fd,3} - \hat{T}_{fd,3}(u_4) \right| \le CI_{\alpha,4}$	$u_4 = [T_{g2}\ T_{fd2}\ x_{fd}]^T$
s_5	$r_5 = \left| x_{bfd} - \hat{x}_{bfd}(u_5) \right| \le CI_{\alpha,5}$	$u_5 = [m_{fd}]$
s_6	$r_6 = \left| T_{g4} - \hat{T}_{g4}(u_6) \right| \le CI_{\alpha,6}$	$u_6 = [T_{g1}\ \theta_{dd}]^T$
s_7	$r_7 = \left| \theta_{dd} - \hat{\theta}_{dd}(u_7) \right| \le CI_{\alpha,7}$	$u_7 = [T_{g1}\ \dot{m}_s\ P_S]^T$
s_8	$r_8 = \left| P_D - \hat{P}_D(u_8) \right| \le CI_{\alpha,8}$	$u_8 = [T_{g1}\ T_{fd3}\ \theta_{dd}]^T$
s_9	$r_9 = \left| H_D - \hat{H}_D(u_9) \right| \le CI_{\alpha,9}$	$u_9 = [T_{g1}\ T_{fd3}\ \theta_{dd}]^T$
s_{10}	$r_{10} = \left| \dot{m}_s - \hat{\dot{m}}_s(u_{10}) \right| \le CI_{\alpha,10}$	$u_{10} = [x_{bfd}\ P_D]^T$
s_{11}	$r_{11} = \left| \dot{m}_{fd} - \hat{\dot{m}}_{fd}(u_{11}) \right| \le CI_{\alpha,11}$	$u_{11} = [x_{bfd}]$
s_{12}	$r_{12} = \left| P_s - \hat{P}_s(u_{12}) \right| \le CI_{\alpha,12}$	$u_{12} = [T_{g1}\ P_D\ \theta_{dd}]^T$

Table H.7 Set of faults in the SAC

Fault name	Process variable
f_1	Chilled water flow rate $\dot{m}_{ch}$
f_2	Chilled water valve percentage opening x_{bfd}
f_3	Temperature of chilled water at the outlet of the SAC T_{ch2}
f_4	Temperature of chilled water at the inlet to the SAC T_{ch1}
f_5	Steam pressure P_s
f_6	Steam flow rate $\dot{m}_s$
f_7	Temperature of cooling water at the outlet of the SAC $T_{co,2}$
f_8	Temperature of cooling water at the inlet to the SAC $T_{co,1}$

Table H.8 Set of residual equations for the SAC and for pth data point

s_j	Fault detection test	Detail of u_j
s_1	$r_1 = \left| \dot{m}_{ch} - \hat{\dot{m}}_{ch}(u_1) \right| \leq CI_{\alpha,1}$	$u_1 = [x_{ch}]^T$
s_2	$r_2 = \left| x_{ch} - \hat{x}_{ch}(u_2) \right| \leq CI_{\alpha,2}$	$u_2 = [\dot{m}_{ch} \ T_{ch1}]^T$
s_3	$r_3 = \left| T_{ch2} - \hat{T}_{ch2}(u_3) \right| \leq CI_{\alpha,3}$	$u_3 = [\dot{m}_{ch} \ T_{ch1}]^T$
s_4	$r_4 = \left| T_{ch1} - \hat{T}_{ch1}(u_4) \right| \leq CI_{\alpha,4}$	$u_4 = [x_{ch} \ T_{ch2} \ \dot{m}_s \ T_{co2}]^T$
s_5	$r_5 = \left| P_s - \hat{P}_s(u_5) \right| \leq CI_{\alpha,5}$	$u_5 = [T_{ch2} \ T_{ch1} \ \dot{m}_s \ T_{co2}]^T$
s_6	$r_6 = \left| \dot{m}_s - \hat{\dot{m}}_s(u_6) \right| \leq CI_{\alpha,6}$	$u_6 = [T_{ch2} \ T_{ch1} \ T_{co2}]^T$
s_7	$r_7 = \left| T_{co2} - T_{co2}(u_7) \right| \leq CI_{\alpha,7}$	$u_7 = [T_{ch2} \ \dot{m}_s]^T$
s_8	$r_8 = \left| T_{co1} - \hat{T}_{co1}(u_8) \right| \leq CI_{\alpha,8}$	$u_8 = [T_{ch2} \ \dot{m}_s]^T$

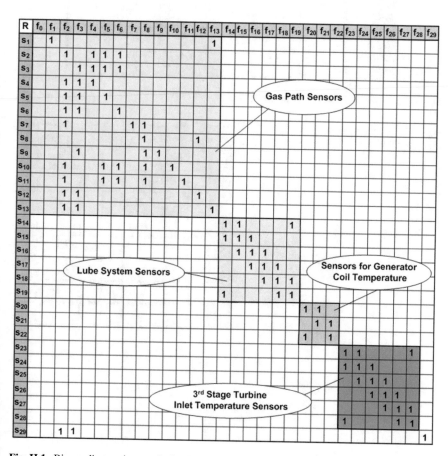

Fig. H.1 Binary diagnostics matrix for GTG

R	f0	f1	f2	f3	f4	f5	f6	f7	f8	f9	f10	f11	f12
S1		NZ			NZ	NZ		NZ					
S2		NZ	NZ		NZ			NZ					
S3				NZ		NZ			NZ				
S4				NZ	NZ	NZ							
S5							NZ					NZ	
S6		NZ					NZ						
S7								NZ			NZ		NZ
S8								NZ	NZ				
S9								NZ		NZ			
S10						NZ			NZ		NZ		
S11						NZ						NZ	
S12								NZ	NZ				NZ

Fig. H.2 Fuzzy diagnostic matrix for the HRSG

Fig. H.3 Fuzzy diagnostics matrix for SAC

R	f0	f1	f2	f3	f4	f5	f6	f7	f8
S1		NZ							
S2		NZ	NZ						
S3			NZ	NZ					
S4					NZ				
S5				NZ		NZ	NZ		
S6							NZ		
S7								NZ	
S8									NZ

Appendix I
ProFDD Main Pages

See Figs. I.1, I.2, I.3, I.4 and I.5.

Fig. I.1 First page of ProFDD

© Springer International Publishing AG 2018
T.A. Lemma, *A Hybrid Approach for Power Plant Fault Diagnostics*,
Studies in Computational Intelligence 743,
https://doi.org/10.1007/978-3-319-71871-2

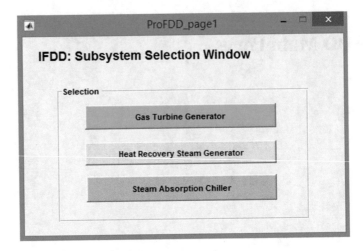

Fig. I.2 Subsystem selection window or menu

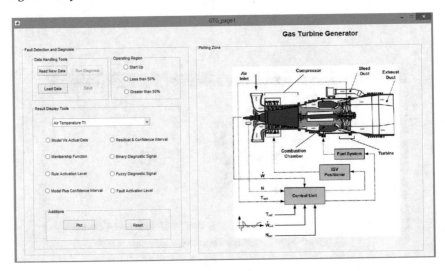

Fig. I.3 Gas turbine generator analysis window

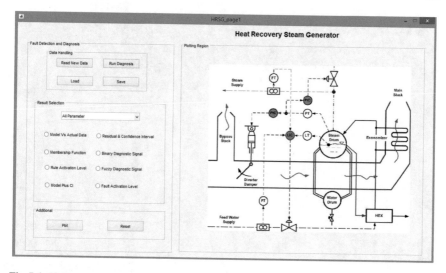

Fig. I.4 Heat recovery steam generator analysis window

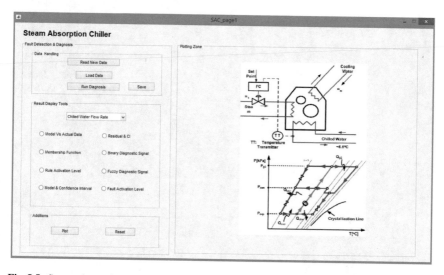

Fig. I.5 Steam absorption chiller analysis window

Printed in the United States
By Bookmasters